ANNALES ACADEMIÆ SCIENTIARUM FENNICÆ

MATHEMATICA

DISSERTATIONES

142

THE BLAGOVEŠČENSKIĬ IDENTITY AND THE INVERSE SCATTERING PROBLEM

KENRICK BINGHAM

Helsinki University of Technology, Department of Engineering Physics and Mathematics

Dissertation for the degree of Doctor of Science in Technology to be presented with due permission of the Department of Engineering Physics and Mathematics for public examination and debate in Auditorium D at Helsinki University of Technology (Espoo, Finland) on the 10th of June, 2005, at 12 noon.

HELSINKI 2005
SUOMALAINEN TIEDEAKATEMIA

ISSN 1239-6303
ISBN 951-41-0974-0

Received 17 May 2005

2000 Mathematics Subject Classification:
Primary 35R30; Secondary 35J10, 35P25, 81U40.

YLIOPISTOPAINO
HELSINKI 2005

Abstract

The inverse scattering problem for the plasma wave equation

$$\left[\partial_t^2 - \triangle + q(x)\right] u(x,t) = 0$$

in three space dimensions is considered in this thesis. It is shown that, under certain assumptions about the potential, the time domain scattering problem can be formulated equivalently in the frequency domain. Time and frequency domain techniques are combined in the subsequent analysis.

The Blagoveščenskiĭ identity is generalised to the case of scattering data, assuming an inverse polynomial decay of the potential. This identity makes it possible to calculate the inner product of certain solutions of the plasma wave equation at a given time, if the corresponding incident waves and the scattering amplitude are known. In the case of a compactly supported potential, these inner products can be calculated for the time derivatives of all solutions.

In the remaining part of the work, the potential is assumed to be compactly supported. A variant of the boundary control method is used to show that using appropriate superpositions of plane waves as incident waves, it is possible to excite a wave basis over a compact set. Letting this set shrink to a point, the Blagoveščenskiĭ identity provides pointwise information about the solutions. When substituted into the plasma wave equation, this yields a method for solving the inverse problem.

Acknowledgements

My work has been supported by the Institute of Mathematics of Helsinki University of Technology, the Academy of Finland, the Finnish Graduate School of Inverse Problems, The Vilho, Yrjö and Kalle Väisälä fund of the Finnish Academy of Science and Letters, the Finnish Foundation for Economic and Technology Sciences – KAUTE and the Foundation of Technology (Tekniikan edistämissäätiö), which I acknowledge with thanks.

I wish to express my sincere gratitude to my advisor and supervisor Prof. Matti Lassas for his expert guidance and continuous encouragement during the process of preparing my thesis. I am most grateful to my initial supervisor Prof. Erkki Somersalo for creating a vibrant inverse problems research group at the Institute of Mathematics of Helsinki University of Technology. I am deeply thankful to Prof. Olavi Nevanlinna for providing excellent working conditions in an appreciative, diversified and pluralistic working environment at the Institute of Mathematics.

It has been a privilege to have Dr. Alexander Kachalov and Doc. Petri Ola, two distinguished experts in the field, as the preliminary examiners of my manuscript. They provided helpful comments and suggested several useful amendments to the text. I am indebted to them for this, as well as for the great alacrity with which they acted.

Based on the very pleasant and fruitful encounters that I already have had with Prof. Yaroslav V. Kurylev, I much look forward to defending my thesis against him and thank him warmly for agreeing to be my opponent.

I am grateful to Prof. Lassi Päivärinta, Dr. Alexander Pushnitski and Dr. Valery Serov for their willingness to discuss questions related to my thesis and to show me how to carry out a few of the calculations that I needed.

I thank all of my teachers and colleagues at the Institute of Mathematics of Helsinki University of Technology and other Finnish inverse problems research groups for their education, help and companionship over the years and for giving me the feeling that I have found my place.

For an understanding of the value of both traditional and emotional intelligence, both of which have been very useful in completing this task, as well as for any skills that I may have acquired in these areas, I am much obliged to my parents. I want to extend my warmest thanks to my beloved wife Ella for her constant support during my work on my thesis, and for letting me realise again and again that the crystalline, abstract beauty of mathematics, as wonderful as it really is, suddenly pales in face of that which is even more important.

Contents

1 Introduction 5
1.1 Time domain scattering . . . 7
1.2 Frequency domain scattering . . . 10
1.3 Notation . . . 14

2 Equivalent formulations of the direct problems 17
2.1 Norm estimates for the solution and its k derivatives . . . 17
2.2 Regularity of solutions . . . 31
2.3 Wave operator . . . 31
2.4 Equivalence result . . . 39

3 Calculating inner products of solutions from scattering data 46
3.1 The one-dimensional case . . . 47
3.2 The Blagoveščenskiĭ identity . . . 48
3.3 Extended far field expansion . . . 51
3.4 Sufficiency of regular scattering data . . . 55
3.5 Scattering amplitude . . . 61
3.6 Inner products of time derivatives . . . 63

4 Scattering control 64
4.1 Simulating point sources . . . 65
4.2 Scattering control . . . 74

5 Solving the inverse problem 77
5.1 Localization . . . 77
5.2 Recovery of the potential . . . 80
5.3 Prospects . . . 81

References 82

1 Introduction

The wave equation with potential

$$\partial_t^2 u(x,t) - \triangle u(x,t) + q(x)u(x,t) = 0, \qquad x \in \mathbb{R}^n, t \in \mathbb{R} \tag{1.1}$$

is known as the *plasma wave equation*, since it models the propagation of electromagnetic waves in low-density plasma under certain conditions, or in the ionosphere [Bal72, Bud61, DR85, JA79, New85]. One of the most elementary models of classical mechanics modelled by (1.1) is the propagation of waves in an elastic medium [MF53].

The plasma wave equation is closely related to the quantum-mechanical Schrödinger equation through the Fourier transform with respect to time [New85, RDC85]. Alternatively, it would be possible to start with the frequency domain equation, which also models acoustic scattering [CK98], and view the time domain wave equation formulation as a tool for analysing it.

Perhaps more importantly for a mathematician, however, (1.1) is one of the simplest non-trivial perturbations to the wave equation, and thus a good starting point for the analysis of the inverse scattering problem for linear hyperbolic second order partial differential equations.

The *direct scattering problem* is, given the real-valued potential q and an incident free space wave, to find the solution of (1.1) that asymptotically coincides with the incident wave in the distant past. The *inverse scattering problem* consists of determining the potential function q from some measurement data of the solutions. In our case, the data will be the full frequency domain scattering amplitude for all directions and all frequencies.

There are other types of inverse scattering problems, too; in particular those of scattering from obstacles and electromagnetic scattering. Much of their theory is similar to that of scattering from a potential [PS02]. The inverse scattering problem also has close connections to various other inverse problems, including inverse problems in bounded domains for the Schrödinger and conductivity equations [Cal80, SU87, Nac88, AP03]. We shall not consider these questions in this study.

We shall restrict ourselves to the three-dimensional case, which may be seen as the most relevant for real world situations. In many respects, the two-dimensional and the general case are similar to the three-dimensional one, but in some crucial places, the 3D case is simpler to handle. In particular, Lax-Phillips scattering theory is simpler in an odd number of space dimensions; this is related to Huygens' principle, which states that when the number of space dimensions is odd, wave fronts in empty space cease to affect a point when they have travelled past it. The fundamental solution of the Helmholtz equation also has useful properties in three dimensions.

A straightforward approach to solving the inverse problem would be to simply calculate the potential from the plasma wave equation:

$$q(x) = \frac{\triangle u(x,t) - \partial_t^2 u(x,t)}{u(x,t)}. \tag{1.2}$$

This formula, however, immediately gives rise to two questions:

1. How do we determine the values of u from the scattering data?

2. How can we make sure that the denominator does not vanish?

A large part of this thesis will be devoted to answering these questions.

The text is organised as follows: The remaining part of Section 1 introduces the direct and inverse scattering problems in more detail, and sets out some notation. In Section 2, we derive a few estimates for the solutions of (1.1) and related equations, and show that the direct problem can be formulated equivalently in the frequency domain, and further as the Lippmann-Schwinger integral equation or through the wave operator. In Section 3, we derive a variant of the *Blagoveščenskiĭ identity*, which allows us to calculate inner products of solutions of (1.1), at any fixed time t_0, from scattering data. If we choose these solutions in such a way that they are supported in a small neighbourhood of a point $x_0 \in \mathbb{R}^3$, these inner products will give information about the behaviour of u near (x_0, t_0), answering Question 1 above. The fact that this is possible is shown in Section 4 for the case of a compactly supported potential, using a variant of the boundary bontrol (BC) method [Bel90, BK92b, Bel97], which we call *scattering control*, as there is now no boundary, but instead, control is done using solutions of the scattering problem. This will also provide an answer to Question 2 above. Varying x_0 and t_0, we can find u, and eventually solve the potential q as in (1.2); this will be done in detail in Section 5.

Although the method we present gives formulae for the reconstruction of the potential, it involves passing to the limit many times and analytic continuation. For this reason, it may not be feasible for practical reconstruction.

When deriving the Blagoveščenskiĭ identity, it is sufficient to assume that the potential q and its first derivatives are real-valued and bounded, that they decay at a certain inverse polynomial rate, and that the corresponding Fredholm operator is injective at zero frequency. When proving the control property and solving the inverse problem, however, we also assume the potential to be compactly supported and once continuously differentiable, and that there are no bound states, *i.e.*, negative eigenvalues of the Schrödinger operator $-\triangle + q$.

The unique solvability of our inverse scattering problem has already been known for some time [Fad56], also for single frequency data [NK87, Nov88, Ram87, Ram88, Ram89, SU87] and for certain classes of non-compactly supported potentials [Nov94, ER95]. The present study, however, provides a novel approach that lends itself to generalization in several directions.

Firstly, it might be possible to relax the assumptions and to cover the case of non-compactly supported potentials as well, since one of the principal tools, the Blagoveščenskiĭ identity, does not depend on this assumption. Our problem is also formally strongly overdetermined, with 5-dimensional data (scattering amplitude $A : S^2 \times S^2 \times \mathbb{R} \to \mathbb{C}$) and a 3-dimensional unknown ($q : \mathbb{R}^3 \to \mathbb{R}$). In many cases it is known that fixed energy data (which are $(2n-2)$-dimensional) determine the n-dimensional potential uniquely, and fewer data might thus suffice here, too.

Secondly, our principal tools — the Blagoveščenskiĭ identity and the control method — are quite geometrical in nature. It would be interesting to attempt to extend the results obtained here to more general geometries and equations. The prospects for this appear promising, as earlier variants of these tools have been developed and applied in such contexts [KKL01]. Already in their present form, our version of the Blagoveščenskiĭ identity may prove useful in other applications.

Scattering can also be viewed in the differential geometric framework [Mel94, Mel95]. By compactifying the manifold, the far field can be transformed into boundary values of the solution, the radiation field [Fri80, Fri01]. Unique continuation from the boundary can then be used to solve the inverse problem if the metric is not too singular at the boundary. This technique was used by Sá Barreto to solve the inverse scattering problem for an asymptotically hyperbolic manifold [SB, Ali84]. An asymptotically Euclidean manifold, however, may be too singular for this method, and the present method could prove to be useful [SB03].

The main contributions of this study are:

1. the generalization of the Blagoveščenskiĭ identity to the scattering case (Theorems 3.11 and 3.12)
2. the introduction of sources simulated by scattered waves (Theorem 4.8) and their use to show the scattering control property (Theorem 4.10)
3. as an application of the Blagoveščenskiĭ identity and scattering control, the derivation of a new reconstruction method for a compactly supported potential which could be developed further to more general settings (Theorem 5.4).

A substantial amount of technical work was also needed for relating the time and frequency domain formulations for scattering to each other, since the spaces that are natural for one formulation are not as natural for another. All calculations for which references are not given were done independently of the existing literature, but most of the results are probably not new, in particular those in Section 2.

1.1 Time domain scattering

Consider the scattering problem, where a free space wave u_i is sent in. Here "free space" means that u_i solves the *wave equation* without potential,

$$\partial_t^2 u_\mathrm{i}(x,t) - \triangle u_\mathrm{i}(x,t) = 0 \tag{1.3}$$

for all $x \in \mathbb{R}^3$, $t \in \mathbb{R}$. A special class of incident waves are the *eventually incoming*, or more precisely *a-incoming*, waves, which are those satisfying the condition (see Figure 1)

$$u_\mathrm{i}(x,t) = 0 \qquad \text{when } |x| < a - t. \tag{1.4}$$

The *Cauchy data*, *i.e.*, the values of a free space solution and its time derivative at any in-

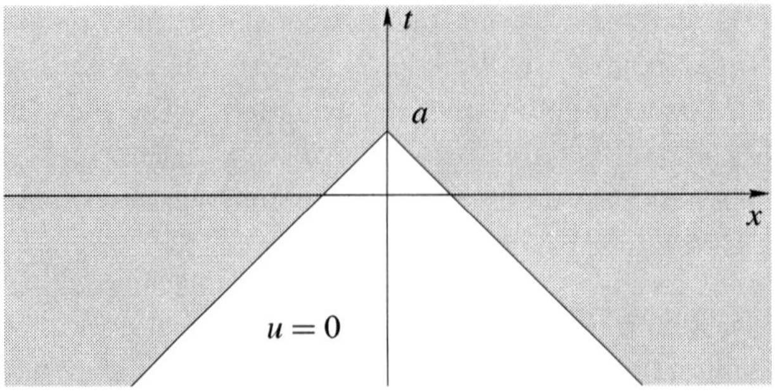

Figure 1: Incoming solution satisfying (1.4).

stant, contain all information about the solution: if the data are known, the wave equation (1.3) can be solved with these initial conditions. These Cauchy data will be called a *wave* at time t.

Lax-Phillips scattering theory [LP67] tells us that each free space solution u_{i}, whose gradient and time derivative are square integrable at any time, has a unique *translation representation* $h \in L^2(\mathbb{R} \times S^2)$, given by the formula

$$\begin{aligned} h(s,\omega) &= \frac{1}{8\pi^2}\left[\partial_s \int_{x\cdot\omega=s+t} \partial_t u_{\mathrm{i}}(x,t)\,dS(x) - \partial_s^2 \int_{x\cdot\omega=s+t} u_{\mathrm{i}}(x,t)\,dS(x)\right] \\ &= \frac{1}{8\pi^2}\left[\partial_s R\left[\partial_t u_{\mathrm{i}}(\cdot,t)\right](s+t,\omega) - \partial_s^2 R\left[u_{\mathrm{i}}(\cdot,t)\right](s+t,\omega)\right] \end{aligned} \tag{1.5}$$

for any $t \in \mathbb{R}$. Here and later on, dS is the standard surface measure on the sphere, and

$$Rf(s,\omega) = \int_{x\cdot\omega=s} f(x)\,dS(x)$$

is the *Radon transform* [Hel99]. The wave is given in terms of its translation representation as

$$\begin{aligned} u_{\mathrm{i}}(x,t) &= \int_{S^2} h(x\cdot\omega - t,\omega)\,dS(\omega) \\ \partial_t u_{\mathrm{i}}(x,t) &= \int_{S^2} \partial_s h(x\cdot\omega - t,\omega)\,dS(\omega); \end{aligned} \tag{1.6}$$

this explains the name. Clearly if h is supported in $(-\infty,-a] \times S^2$, the wave satisfies the a-incoming condition (1.4). Conversely, if $\operatorname{supp} h \subset [-b,\infty) \times S^2$, the wave is *eventually outgoing*, or *b-outgoing*, *i.e.*, $u(x,t) = 0$ when $|x| < t+b$. Actually for $a = 0$, this is also a necessary condition [Hel99, Corollary I.7.4].

Now since there is a potential, the incident wave u_{i} does not solve the plasma wave equation (1.1), but if it is corrected by a suitable scattered wave u_{s}, the total wave $u = u_{\mathrm{i}} + u_{\mathrm{s}}$ may be a solution. Thinking of this scattered wave as physically arising from

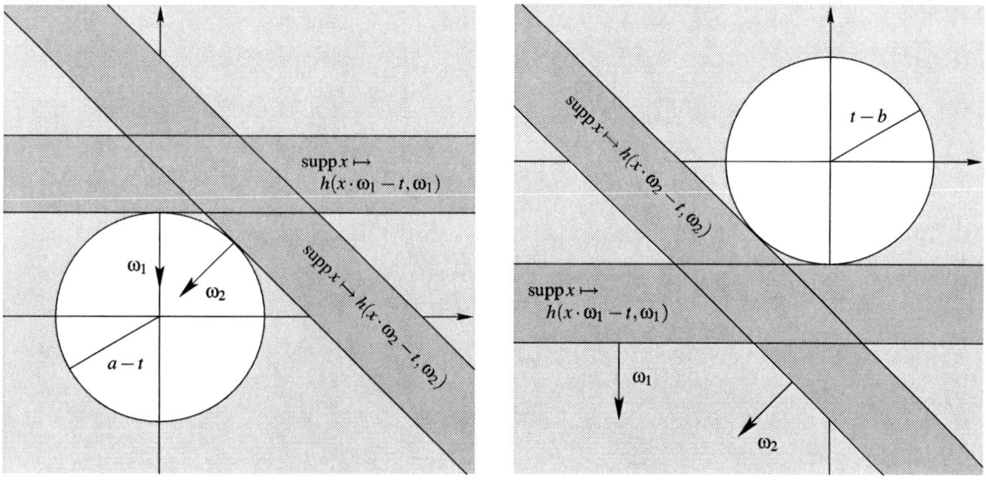

Figure 2: An a-incoming (left) and b-outgoing (right) wave as a superposition of plane waves.

interaction of the wave with the potential, it is natural to require it to be *causal*. By causality, we mean roughly speaking that u_s may only depend on the past: the plasma wave equation (1.1) can be written

$$\left[\partial_t^2 - \Delta + q(x)\right] u_s(x,t) = -q(x) u_i(x,t),$$

and we say that u_s is causal if $u_s(x_0,t_0)$ depends on the right hand side only in the backward light cone

$$\left\{(x,t) \in \mathbb{R}^3 \times \mathbb{R} \,\middle|\, |x - x_0| \leq t_0 - t\right\}.$$

A precise definition is given in terms of the advanced fundamental solution of the wave equation: u_s is *causal* if and only if

$$\begin{aligned} u_s(x_0,t_0) &= -\int_{-\infty}^{t_0} \frac{1}{4\pi(t_0 - t)} \int_{|x-x_0|=t_0-t} q(x) u(x,t)\, dS(x)\, dt \\ &=: -\left[E_+ * (qu)\right](x_0,t_0). \end{aligned} \tag{1.7}$$

This convolution makes sense if q is compactly supported and u_i (and thus u) is incoming: then $\operatorname{supp} qu \subset [-M,\infty) \times \overline{B_M}$ for some $M > 0$, which together with the fact that $\operatorname{supp} E_+ \subset [0,\infty) \times \mathbb{R}^3$ yields that the mapping

$$\operatorname{supp} E_+ \times \operatorname{supp}(qu) \ni \big((x,t),(y,s)\big) \mapsto (x+y, s+t) \in \mathbb{R}^{3+1}$$

is proper [Hör90, p. 104]. Also if q decays fast enough, this formal convolution converges for sufficiently quickly decaying u; this will be shown in Theorem 2.24.

The frequency domain analogue of the the integral equation (1.7), which will shortly be introduced, is known as the Lippmann-Schwinger equation. For this reason, we call (1.7) the *time domain Lippmann-Schwinger equation* and formulate the following:

Time domain direct scattering problem: *Given the potential q and an incident wave u_{i}, determine a causal scattered wave u_{s} such that $u = u_{\mathrm{i}} + u_{\mathrm{s}}$ satisfies the time domain Lippmann-Schwinger equation (1.7).*

Another time-domain formulation for the scattering problem is in terms of the *wave operator* [LP67, RS79]

$$\Omega^{\pm} = \lim_{t \to \mp\infty} W_1(-t) W_0(t),$$

where W_0 and W_1 are the *propagators*

$$W_0(t) : \begin{pmatrix} u_{\mathrm{i}}(\cdot, s) \\ \partial_t u_{\mathrm{i}}(\cdot, s) \end{pmatrix} \mapsto \begin{pmatrix} u_{\mathrm{i}}(\cdot, s+t) \\ \partial_t u_{\mathrm{i}}(\cdot, s+t) \end{pmatrix} \qquad W_1(t) : \begin{pmatrix} u(\cdot, s) \\ \partial_t u(\cdot, s) \end{pmatrix} \mapsto \begin{pmatrix} u(\cdot, s+t) \\ \partial_t u(\cdot, s+t) \end{pmatrix},$$

the definition being independent of $s \in \mathbb{R}$. This operator and its relation to the integral equation formulation above will be investigated in Section 2.3.

1.2 Frequency domain scattering

The Fourier transform with respect to time of any solution u of (1.1) is defined for integrable functions by[1]

$$\hat{u}(x,k) = \mathcal{F}\big(u(x,\cdot)\big)(k) = \int_{-\infty}^{\infty} e^{ikt} u(x,t)\, dt,$$

and for tempered distributions $u \in \mathcal{S}'$ by duality, as usual. Assuming that $u \in \mathcal{S}'$, we see that $\hat{u}$ clearly solves, in the sense of distributions, the frequency domain plasma wave equation, or the Schrödinger eigenvalue problem[2]

$$\big(-\triangle - k^2 + q(x)\big)\hat{u}(x,k) = 0, \qquad x \in \mathbb{R}^3, \tag{1.8}$$

[1] We make this, somewhat less common choice of the plus sign in the exponent. This allows us to keep the time domain solution u as the starting point and still get the usual signs in the Sommerfeld radiation condition (1.10). The inverse Fourier transform is then, for integrable functions,

$$\check{f}(x,t)u(x,t) = \mathcal{F}^{-1}\big(f(x,\cdot)\big)(t) = \frac{1}{2\pi}\int_{-\infty}^{\infty} e^{-ikt} f(x,k)\, dk.$$

[2] If we now make the inverse Fourier transform with respect to the variable k^2 instead of k, we arrive at the time-dependent Schrödinger equation

$$\left[-i\frac{\partial}{\partial t} - \triangle + q\right]\psi = 0.$$

for all wave numbers $k \in \mathbb{R}$. This frequency domain formulation is often taken as the starting point because of its significance in quantum-mechanical scattering. When dealing entirely in the frequency domain, the wave number k is usually viewed as fixed, corresponding to a time harmonic wave with a single frequency $k/2\pi$. We shall now, however, combine frequency and time domain techniques, and therefore need all frequencies $k \in \mathbb{R}$.

Most results about frequency domain scattering mentioned below can be found in [CK98] for classical solutions and compactly supported potentials q. Some generalizations are derived later in this thesis.

The Fourier transform of the incoming wave $\hat{u}_{\mathrm{i}}(x,k) = \mathcal{F}(u_{\mathrm{i}}(x,\cdot))(t)$ clearly satisfies the *Helmholtz equation*

$$(-\triangle - k^2)\hat{u}_{\mathrm{i}}(x,k) = 0, \qquad x \in \mathbb{R}^3. \tag{1.9}$$

In the frequency domain, the direct scattering problem thus becomes: Given the potential q and the incident field $\hat{u}_{\mathrm{i}}$ satisfying (1.9), find the scattered field $\hat{u}_{\mathrm{s}}$ such that $\hat{u} = \hat{u}_{\mathrm{i}} + \hat{u}_{\mathrm{s}}$ satisfies (1.8). The causality condition of $\hat{u}_{\mathrm{s}}$ translates to the *Sommerfeld radiation condition*

$$\frac{\partial \hat{u}_{\mathrm{s}}}{\partial r} - ik\hat{u}_{\mathrm{s}} = o\left(\frac{1}{r}\right) \qquad \text{as } r := |x| \to \infty \tag{1.10}$$

uniformly in all directions of $\hat{x} := \frac{x}{r}$.

Frequency domain direct scattering problem: *Given the potential q and a solution $\hat{u}_{\mathrm{i}}$ of (1.9), find $\hat{u}_{\mathrm{s}}$ such that $\hat{u} = \hat{u}_{\mathrm{i}} + \hat{u}_{\mathrm{s}}$ satisfies (1.8) and the Sommerfeld radiation condition (1.10).*

The integral equation formulation for this problem is the *Lippmann-Schwinger equation*:

$$\hat{u}_{\mathrm{s}}(\cdot,k) = -\mathcal{G}_k\big(q\hat{u}(\cdot,k)\big), \tag{1.11}$$

where $\mathcal{G}_k\varphi = \Phi(\cdot,k) * \varphi$ and

$$\Phi(x,k) := \hat{E}_+(x,k) = \frac{e^{ik|x|}}{4\pi|x|}$$

is the radiating fundamental solution to the Helmholtz equation, *i.e.*, $-(\triangle + k^2)\Phi = \delta$ and Φ satisfies the Sommerfeld radiation condition. The convolution in (1.11) is taken with respect to the space variable $x \in \mathbb{R}^3$ only.

The equivalence of the time domain direct scattering problem and these two frequency domain direct scattering problems is shown at the end of Section 2.

The Sommerfeld radiation condition explains why the solution $\hat{u}(x,k)$ is written as a function of k and not k^2 or $|k|$: even though Equation (1.8) is the same for k and $-k$, the Sommerfeld radiation condition (1.10) is not. However, when the potential is real valued, solutions for negative k are essentially redundant: If $\hat{u}(x,k)$ is known for $k \geq 0$, $\hat{u}(x,-k) = \overline{\hat{u}(x,k)}$ gives the solution to (1.8), (1.10) for $-k \leq 0$.

Solutions satisfying the Sommerfeld radiation condition have the behaviour

$$\hat{u}_{\mathrm{s}}(x,k) = \frac{e^{ik|x|}}{|x|}\hat{u}_{\mathrm{s}}^{\infty}(\hat{x},k) + o\left(\frac{1}{|x|}\right) \qquad \text{as } |x| \to \infty,$$

where $\hat{u}_{\mathrm{s}}^{\infty}$ is the *far field pattern*. In the special case where the potential is compactly supported and

$$\hat{u}_{\mathrm{s}}(x,k) = \frac{e^{ik|x|}}{|x|} f(k)$$

for large $|x|$, we have an outgoing spherical wave

$$u_{\mathrm{s}}(x,t) = \frac{\check{f}(t-|x|)}{|x|},$$

which is a natural example of an outgoing wave in the time domain. Assuming the k dependence of the remainder term in the Sommerfeld radiation condition to be integrable, the inverse Fourier transform gives

$$\partial_r u_{\mathrm{s}}(x,t) + \partial_t u_{\mathrm{s}}(x,t) = o\left(\frac{1}{r}\right) \qquad \text{as } r \to \infty,$$

uniformly in all directions and all $t \in \mathbb{R}$. This condition is of course satisfied by the outgoing spherical wave, assuming some regularity of $\check{f}$:

$$(\partial_r + \partial_t)\frac{\check{f}(t-|x|)}{|x|} = -\frac{\check{f}(t-|x|)}{|x|^2} - \frac{\check{f}'(t-|x|)}{|x|} + \frac{\check{f}'(t-|x|)}{|x|} = -\frac{\check{f}(t-|x|)}{|x|^2}.$$

The *limiting absorption principle* tells us that, instead of real k in the Lippmann-Schwinger equation (1.11), we can consider the limit from the complex upper half plane:

$$\mathcal{G}_k = \lim_{\varepsilon \searrow 0} \mathcal{G}_{k+i\varepsilon}.$$

For $\varepsilon > 0$, the operator $\mathcal{G}_{k+i\varepsilon}$ models a physical situation where absorption occurs in addition to scattering. The limiting absorption principle now says that the no absorption case is the limit of cases of weaker and weaker absorption. [Agm75]

Denote by F the Fourier transform with respect to $x \in \mathbb{R}^3$, defined for integrable functions by[3]

$$Ff(\xi) = \int_{\mathbb{R}^3} e^{-ix\cdot\xi} f(x)\, dx$$

[3]Now the sign in the exponent is the more usual one, in contrast to to definition of $\mathcal{F}$. The inverse Fourier transform is now for integrable functions

$$F^{-1}f(\xi) = \frac{1}{(2\pi)^n}\int_{\mathbb{R}^3} e^{ix\cdot\xi} f(\xi)\, d\xi.$$

and for $f \in \mathcal{S}'$ again by duality. For $\varepsilon > 0$, the L^p convolution theorem tells that F transforms the convolution by Φ in the Lippman-Schwinger equation into multiplication by its Fourier transform, which will be calculated in Lemma 2.8. For the case of no absorption, we must take the limit:

$$[F(\Phi * f)](\xi) = \lim_{\varepsilon \searrow 0} \frac{(Ff)(\xi)}{\xi^2 - (k+i\varepsilon)^2} =: \frac{(Ff)(\xi)}{\xi^2 - (k+i0)^2}.$$

The *inverse scattering problem* is to determine the unknown potential q, when some information about the scattered wave corresponding to each incoming wave u_{i} is known. This information is in our case the *scattering data, i.e.*, the far field patterns

$$\hat{u}_{\mathrm{s}}^{\infty}(\hat{x},k) = \lim_{r\to\infty} re^{-ikr}\hat{u}_{\mathrm{s}}(r\hat{x},k), \qquad \hat{x} \in S^2, k \in \mathbb{R},$$

corresponding to different incident waves u_{i}. More precisely, we use the *scattering amplitude* A, which is the far field pattern corresponding to an incident plane wave: $A(\omega,\hat{x};k) = \hat{v}_{\mathrm{s}}^{\infty}(\hat{x},k;\omega)$, where

$$\begin{aligned} \left[-\triangle - k^2 + q(x)\right]\hat{v}(x,k;\omega) &= 0 \\ \hat{v}(x,k;\omega) &= \hat{v}_{\mathrm{s}}(x,k;\omega) + \hat{v}_{\mathrm{i}}(x,k;\omega) \\ \hat{v}_{\mathrm{i}}(x,k;\omega) &= e^{ik\hat{x}\cdot\omega} \\ \frac{\partial \hat{v}_{\mathrm{s}}(r\hat{x},k;\omega}{\partial r} - ik\hat{v}_{\mathrm{s}}(x,k;\omega) &= o\left(\frac{1}{r}\right) \qquad \text{as } r := |x| \to \infty, \text{ uniformly over all } \hat{x} \in S^2. \end{aligned}$$

The Fourier transform of the expression (1.6) of a free space wave in terms of its translation representation yields a *Herglotz wave function, i.e.*, a linear combination of plane waves

$$\hat{u}_{\mathrm{i}}(x,k) = \int_{S^2} e^{ikx\cdot\omega}\hat{h}(k,\omega)\,dS(\omega),$$

where $\hat{h} \in L^2(\mathbb{R}\times S^2)$.

In addition to the far field, we shall also use a few lower order terms of the *extended far field expansions*

$$P\hat{u}_{\mathrm{s}}(r\hat{x},k) = e^{ikr}\sum_{j=j_0^P}^{N} \frac{U_j^P(\hat{x},k)}{r^j} + o\left(\frac{1}{r^N}\right), \quad \hat{x} \in S^2, k \in \mathbb{R}, P \in \{1, \partial_r, \partial_k^2, \partial_r\partial_k^2\}.$$

The mapping of the incident wave u_{i} to the coefficients U_j^P is called the *extended scattering data*. These data will be only be a tool in an intermediate stage, for it turns out that under appropriate conditions, the usual far field $\hat{u}_{\mathrm{s}}^{\infty} = U_1^1$ determines the lower order coefficients.

The rest of this study will be devoted to developing tools for solving the inverse problem.

1.3 Notation

In addition to the more standard notation and that already set out, we use the following. We write $A := B$ for A being defined as B. We also write

$$\mathbb{C}_+ := \{z \in \mathbb{C} \,|\, \operatorname{Im} z > 0\} \qquad \text{upper half space}$$

$$1_U(x) := \begin{cases} 1, & \text{when } x \in U \\ 0, & \text{when } x \notin U. \end{cases} \qquad \text{characteristic function}$$

$$B(x_0, r) := \{x \in \mathbb{R}^n \,|\, |x - x_0| < r\},\ B_r := B(0, r) \qquad \text{open balls}$$

$$S^{n-1}(x_0, r) := \{x \in \mathbb{R}^n \,|\, |x - x_0| = r\},\ S^{n-1} := S^{n-1}(0, 1) \qquad \text{spheres.}$$

Multi-indices $\alpha \in \mathbb{N}^n$ ($\mathbb{N} = \{0, 1, 2, \ldots\}$) are used often, with

$$\partial^\alpha = \frac{\partial^{|\alpha|}}{\partial x_1^{\alpha_1} \cdots \partial x_n^{\alpha_n}}, \qquad |\alpha| = \alpha_1 + \cdots + \alpha_n.$$

We denote the unit vector in $\mathbb{N}^n \subset \mathbb{R}^n$ by $e_j = (0, \ldots, 0, 1, 0, \ldots, 0)$ with 1 in the jth place.

When estimating different quantities, we use the letter C to denote constants. Its value may change from occurrence to occurrence, even within the same formula.

The evaluation of a distribution f with a test function g over $X \subset \mathbb{R}^n$ is written $\langle f, g \rangle$. We sometimes also slightly abuse notation and write distributions with variables as in $\delta_{x_0}(y)$, in analogy with locally integrable functions, and write formal integrals

$$\langle f(x,y), g(x,y) \rangle_x := \langle f(\cdot, y), g(\cdot, y) \rangle =: \int_X f(x,y) g(x,y)\, dx. \tag{1.12}$$

This simplifies the notation when working in product spaces, as does $\|f(s)\|_{X(s)} := \|s \mapsto f(s)\|_X$. The "prototype" variable in $\mathbb{R}^n$ is $x = r\hat{x}$ with $r = |x|$ and $\hat{x} = x/r$. With a prototype variable we mean one that we may sometimes introduce even if it has been omitted in earlier stages of the calculations; also differentiations can be written with respect to this variable. In $\mathbb{R}$ the prototype variable is t, in particular when referring to time. After the Fourier transform with respect to x and t, the prototype variables are ξ and k, respectively.

The reflection operator with respect to time is written $(\mathcal{R}f)(x,t) = f(x, -t)$. The projection operator is written $\pi_j : A_1 \times \cdots \times A_n \ni (x_1, \ldots, x_n) \mapsto x_j \in A_j$ or sometimes π_t when t is the prototype variable of one of the factor spaces A_j. The image of a set $U \subset X$ under an operator $T : X \to Y$ is written $TU = \{Tx \in Y \,|\, x \in U\}$. The Banach space of bounded linear operators from a Banach space X to a Banach space Y is written $\mathcal{L}(X, Y)$.

The inner product in a Hilbert space $\mathcal{H}$ is written $(f, g)_{\mathcal{H}}$. Most often we have $\mathcal{H} = L^2(X)$ with $X \subset \mathbb{R}^n$ and

$$(f, g) := (f, g)_{L^2(X)} = \int_X f(x) \overline{g(x)}\, dx. \tag{1.13}$$

The notations (1.12) and (1.13) are also used for the respective right hand sides whenever f and g are functions whose product is integrable. The Hölder conjugate exponent of $p \in [1, \infty]$ is denoted by p'; this is the number for which $1/p + 1/p' = 1$.

1.3.1 Weight functions

We define the *weight function*

$$w(x) = \sqrt{1+|x|^2} \qquad \text{with } x \in \mathbb{R}^n \text{ for any } n \in \mathbb{Z}_+. \tag{1.14}$$

The notation $\langle x \rangle$ is used in many texts; we choose to write $w(x)$ to simplify some notation. The weight function will appear, for instance, in the weighted L^p norm

$$\|f\|_{L^p_\delta(X)} := \|w^\delta f\|_{L^p(X)},$$

for $\delta \in \mathbb{R}$, $1 \le p \le \infty$ and measurable $X \subset \mathbb{R}^n$. Obviously, we define

$$L^p_\delta(X) = \left\{ f \in L^1_{\text{loc}}(X) \,\middle|\, \|f\|_{L^p_\delta(X)} < \infty \right\}.$$

Clearly

$$\|1\|_{L^p_\rho(\mathbb{R}^n)} = \|w^\rho\|_{L^p(\mathbb{R}^n)} < \infty \qquad \text{if } 1 \le p < \infty \text{ and } \rho < -n/p. \tag{1.15}$$

The precise form of the weight function usually does not matter, and we shall sometimes use the equivalent forms

$$1+|x| \sim \max\{1,|x|\} \sim w(x) \qquad w(x)^\rho \sim w(|x|^\rho) \quad \text{when } \rho \ge 0, \tag{1.16}$$

which are easier to handle in some situations. We choose the form (1.14) as the definition, because it is also smooth at the origin. This will give more flexibility when working with weighted Sobolev spaces in Section 2.1. There we shall need the following estimate, which would not hold, had we chosen $w(x) = 1+|x|$.

Lemma 1.1. *Let* $\alpha \in \mathbb{N}^n$, $\delta \in \mathbb{R}$. *Then*

$$\partial^\alpha w(x)^\delta = \sum_{j=0}^{|\alpha|} \sum_{|\beta| \le 2j-|\alpha|} c_{\alpha,\beta,\delta,j,n} w(x)^{\delta-2j} x^\beta$$

for some constants $c_{\alpha,\beta,\delta,j,n} \in \mathbb{R}$. *In particular,*

$$|\partial^\alpha w(x)^\delta| \le C_{\alpha,\delta,n} w(x)^{\delta-|\alpha|}. \tag{1.17}$$

Proof. By induction: The claim is trivially true for $|\alpha| = 0$. Assume that it is true for ∂^a, $a \in \mathbb{N}^n$, and let $\alpha = a + e_m$. Then

$$\partial^\alpha w(x)^\delta = \partial_m \partial^a w(x)^\delta = \sum_{j=0}^{|a|} \sum_{|\beta| \le 2j-|a|} c_{a,\beta,\delta,j,n} \partial_m \left[\left(1+|x|^2\right)^{\frac{\delta}{2}-j} x^\beta \right].$$

Now

$$\begin{aligned}
\partial_m \left[(1+|x|^2)^{\frac{\delta}{2}-j} x^\beta \right] &= (\delta-2j) \left(1+|x|^2\right)^{\frac{\delta}{2}-(j+1)} x_m x^\beta + \left(1+|x|^2\right)^{\frac{\delta}{2}-j} \partial_m x^\beta \\
&= C_1 w(x)^{\delta-2(j+1)} x^{\beta+e_m} + C_2 w(x)^{\delta-2j} x^{\beta-e_m},
\end{aligned}$$

with $C_2 = 0$ if $\beta_m = 0$. These terms are of the required form, since $j+1 \le |a|+1 = |\alpha|$ and $|\beta \pm e_m| \le |\beta|+1 \le 2j-|a|+1 = 2(j+1)-|\alpha|$. The estimate (1.17) follows from the observation

$$\left|w(x)^{\delta-2j}x^{\beta}\right| \le w(x)^{\delta-2j}|x|^{|\beta|} \le w(x)^{\delta-2j+|\beta|} \le w(x)^{\delta-|\alpha|}.$$

□

1.3.2 Convolution with respect to time

We shall sometimes want to smooth distributions $u \in \mathcal{D}'(\mathbb{R}^3 \times \mathbb{R})$ with respect to time only. This can be done by convolving with respect to the time variable, defined for integrable functions by

$$(\psi *_t \varphi)(x,t) = \int_{\mathbb{R}} \psi(x,t-s)\varphi(s)\,ds,$$

as usual. For distributions $u \in \mathcal{D}'(X \times \mathbb{R})$ and $v \in \mathcal{D}'(Y \times \mathbb{R})$ with v compactly supported with respect to the variable $t \in \mathbb{R}$, we define for $\varphi \in C_0^\infty(X)$, $\psi \in C_0^\infty(Y)$ and $\theta \in C_0^\infty(\mathbb{R})$

$$\langle u *_t v, \varphi \otimes \psi \otimes \theta\rangle = \big\langle u(x,s)v(y,t), \varphi(x)\psi(y)\chi(t)\theta(s+t)\big\rangle_{x,y,s,t} \tag{1.18}$$

where $\chi \in C_0^\infty(\mathbb{R})$ is such that $\chi \equiv 1$ in $\pi_t \operatorname{supp} v$. This defines the distribution $u *_t v$ since test functions of the above tensor product form are dense [Trè67, Thm. 39.2]; this fact will be used repeatedly in what follows. In the case of integrable u and v, these definitions agree: with the change of variable $t = \sigma - s$, $\sigma = s+t$, we get

$$\begin{aligned}\langle u *_t v, \varphi \otimes \psi \otimes \theta\rangle &= \int_X \int_Y \int_{\mathbb{R}} \int_{\mathbb{R}} u(x,s)v(y,\sigma-s)\,ds\,\varphi(x)\psi(y)\theta(\sigma)\,d\sigma\,dy\,dx\\ &= \int_X \int_Y \int_{\mathbb{R}} \int_{\mathbb{R}} u(x,s)v(y,t)\,\varphi(x)\psi(y)\chi(t)\theta(s+t)\,ds\,dt\,dy\,dx.\end{aligned}$$

A convolution with respect to $x \in \mathbb{R}^3$ only will also appear in $\mathcal{G}_k\varphi = \Phi *_x \varphi$. In this case, the frequency variable k is fixed, and the convolution can be viewed as a Lebesgue integral.

2 Equivalent formulations of the direct problems

In this section, we shall show that the time and frequency domain scattering problems and the Lippmann-Schwinger equation

$$(I+\mathcal{G}_k Q)\hat{u}_\mathrm{s} = -\mathcal{G}_k(q\hat{u}_\mathrm{i})$$

are equivalent to each other; Q denotes the operator of pointwise multiplication by the potential q. This equation will be considered in the context of certain weighted L^p spaces, where we can use L. Päivärinta's extension [Päi04] to S. Agmon's classical result [Agm75] telling that the operator $\mathcal{G}_k : \varphi \mapsto \Phi(\cdot,k) * \varphi$ is bounded in these spaces. It is also natural to use the weights because one of the simplest solutions to the Helmholtz equation, the plane wave $e^{ikx\cdot\omega}$, is of constant absolute value and thus does not belong to the unweighted spaces.

Our strategy is the following. We first estimate the norms of the incident part $\hat{u}_\mathrm{i}$, under certain assumptions about the translation representation. Then we show that once we have these estimates, the corresponding norms of

$$\hat{u}_\mathrm{s}(\cdot,k) = -(I+\mathcal{G}_k Q)^{-1}\mathcal{G}_k q\hat{u}_\mathrm{i}(\cdot,k) \tag{2.1}$$

can be estimated by the norms of $\hat{u}_\mathrm{i}$, by proving that the operators $(I+\mathcal{G}_k Q)^{-1}$ and $\mathcal{G}_k Q$ exist and that they and their derivatives with respect to k are bounded. This also establishes the unique solvability of the Lippmann-Schwinger equation.

Even though the connection between the time and frequency domain equations (1.1) and (1.8) is simply the Fourier transform, the function spaces that are natural for the different formulations are not related in the most straightforward manner. For this reason, our proof of the equivalence entails some work. On the other hand, the estimates that will soon be proved will also be useful in what follows, in particular in the proof of the Blagoveščenskiĭ identity in Section 3.

2.1 Norm estimates for the solution and its k derivatives

We shall now estimate weighted L^p norms of $\hat{u}$ and its two first derivatives with respect to k. These estimates are used to deduce regularity properties of the solution $\hat{u} = \hat{u}_\mathrm{i} + \hat{u}_\mathrm{s}$ and furthermore show the equivalence of the frequency domain direct scattering problem and the time and frequency domain Lippmann-Schwinger equations (1.7) and (1.11).

These estimates will also be needed later, for $p = 2$ and some $p > 3$, in the derivation of the extended far field expansions, which are used in the Blagoveščenskiĭ identity of Theorem 3.1. The Sobolev embedding theorems and the estimates for $\mathcal{G}_k$ that we shall be using impose here the additional requirement that $p < 6$.

For all incident waves u_i whose translation representations h are in $L^2(\mathbb{R}\times S^2)$, we have the following weighted L^2 estimate.

Lemma 2.1. *Let*

$$u_{\mathrm{i}}(x,t) = \int_{S^2} h(x\cdot\omega - t,\omega)\, dS(\omega)$$

with $h \in L^2(\mathbb{R}\times S^2)$. *Then*

$$\|\hat{u}_{\mathrm{i}}(x,\cdot)\|_{L^2(\mathbb{R})} = \sqrt{2\pi}\,\|u_{\mathrm{i}}(x,\cdot)\|_{L^2(\mathbb{R})}$$

is uniformly bounded with respect to $x \in \mathbb{R}^3$, *and in particular,*

$$\|\hat{u}_{\mathrm{i}}\|_{L^2_{(-\delta,0)}} := \left[\int_{\mathbb{R}^3}\int_{\mathbb{R}} |\hat{u}_{\mathrm{i}}(x,k)|^2 w(x)^{-2\delta}\, dk\, dx\right]^{1/2} = \sqrt{2\pi}\,\|u_{\mathrm{i}}\|_{L^2_{(-\delta,0)}} < \infty \tag{2.2}$$

for all $\delta > 3/2$.

Proof. Parseval's formula and Hölder's inequality tell us that

$$\begin{aligned}\|\hat{u}_{\mathrm{i}}(x,\cdot)\|^2_{L^2(\mathbb{R})} &= 2\pi\,\|u_{\mathrm{i}}(x,\cdot)\|^2_{L^2(\mathbb{R})}\\ &\le 2\pi\int_{\mathbb{R}}\left|\int_{S^2} |h(x\cdot\omega - t,\omega)|\, dS(\omega)\right|^2 dt\\ &\le 8\pi^2\int_{S^2}\int_{\mathbb{R}} |h(x\cdot\omega - t,\omega)|^2\, dt\, dS(\omega).\end{aligned}$$

Therefore

$$\begin{aligned}\|\hat{u}_{\mathrm{i}}\|^2_{L^2_{(-\delta,0)}(\mathbb{R}^3\times\mathbb{R})} &= \int_{\mathbb{R}^3}\int_{\mathbb{R}} |\hat{u}_{\mathrm{i}}(x,k)|^2\, dk\, w(x)^{-2\delta}\, dx\\ &= 2\pi\int_{\mathbb{R}^3}\int_{\mathbb{R}} |u_{\mathrm{i}}(x,t)|^2\, dt\, w(x)^{-2\delta}\, dx\\ &\le 8\pi^2\|h\|^2_{L^2(\mathbb{R}\times S^2)}\|w^{-2\delta}\|_{L^1(\mathbb{R}^3)} < \infty\end{aligned}$$

by (1.15). □

We shall also need stronger estimates of the k dependence. To obtain these estimates, we restrict ourselves to incident waves whose translation representations are fairly smooth and compactly supported.

Lemma 2.2. *Let*

$$u_{\mathrm{i}}(x,t) = \int_{S^2} h(x\cdot\omega - t,\omega)\, dS(\omega),$$

$h \in C_0^b(\mathbb{R}\times S^2)$, $b \in \mathbb{N}$, $m \in \mathbb{N}$, $\delta > m + 3/p$, $1 \le p < \infty$. *Then for all* $k \in \mathbb{R}$,

$$\left\|\frac{\partial^m \hat{u}_{\mathrm{i}}}{\partial k^m}(\cdot,k)\right\|_{L^p_{-\delta}} \le \frac{C}{|k|^b}$$

for some constant $C = C(\delta,m,p,h)$.

Proof. In the Fourier transform, the factor k is clearly translated into a t derivative, and the k derivative into a factor t:

$$|k|^{pb}\left\|\frac{\partial^m \hat{u}_{\mathrm{i}}}{\partial k^m}(\cdot,k)\right\|_{L^p_{-\delta}}^p = \int_{\mathbb{R}^3}\left|\int_{\mathbb{R}} k^b \partial_k^m e^{ikt} u_{\mathrm{i}}(x,t)\,dt\right|^p w(x)^{-p\delta}\,dx$$
$$= \int_{\mathbb{R}^3}\left|\int_{\mathbb{R}} e^{ikt}\partial_t^b t^m u_{\mathrm{i}}(x,t)\,dt\right|^p w(x)^{-p\delta}\,dx.$$

Now making the change of variables $s = x\cdot\omega - t$ and observing that $|t| \le |x| + |s| \le w(x)w(s)$ we get that

$$\left|\int_{\mathbb{R}} e^{ikt}\partial_t^b t^m u_{\mathrm{i}}(x,t)\,dt\right| \le C\sum_{j=0}^{b}\int_{S^2}\int_{\mathbb{R}} |t|^{m-b+j}|\partial_t^j h(x\cdot\omega - t,\omega)|\,dt\,dS(\omega)$$
$$\le C\sum_{j=0}^{b}\|\partial_s^j h\|_{L^1_m} w(x)^m.$$

Therefore

$$|k|^{pb}\left\|\frac{\partial^m \hat{u}_{\mathrm{i}}}{\partial k^m}(\cdot,k)\right\|_{L^p_{-\delta}}^p \le C\sum_{j=0}^{b}\|\partial_s^j h\|_{L^1_m}^p\,\|w^{m-\delta}\|_{L^p}^p,$$

which is finite by our assumption about δ. □

Now having established estimates for the $L^p_{-\delta}$ norms of the incident field $\hat{u}_{\mathrm{i}}$ and its k derivatives, we do the same for the scattering solutions $\hat{u}_{\mathrm{s}}(\cdot,k)$ of the Lippmann-Schwinger equation (2.1). To this end, we analyse the operators $(I+\mathcal{G}_k Q)^{-1}$ and $\mathcal{G}_k Q$ and their k derivatives.

Lemma 2.3. *Let* $1 \le m \le p < \infty$, *and* $q \in L^\infty_\gamma$ *for some* $\gamma > \delta + \rho + 3\frac{p-m}{pm}$ *and any* $\delta, \rho \in \mathbb{R}$. *Then*

$$\|Qf\|_{L^m_\rho(\mathbb{R}^3)} \le C\|f\|_{L^p_{-\delta}(\mathbb{R}^3)}$$

for some constant C.

Proof. Simply use Hölder's inequality and (1.15):

$$\|Qf\|_{L^m_\rho}^m = \int_{\mathbb{R}^3}|q(x)|^m|f(x)|^m w(x)^{m\rho}\,dx$$
$$\le C\int_{\mathbb{R}^3}|f(x)|^m w(x)^{m(\rho-\gamma)}\,dx$$
$$\le C\,\|f^m w^{-\delta m}\|_{L^{p/m}}^m\,\|w^{m(\rho+\delta-\gamma)}\|_{L^{(p/m)'}}^m \le C\,\|f\|_{L^p_{-\delta}}^m.$$

□

Lemma 2.4. *Let* $1 < m \leq 2 \leq p < \infty$ *and* $\delta > \max\{1, 3/p, 3(1-1/m)\}$. *Then the operator valued function*

$$\mathbb{R} \ni k \mapsto \mathcal{G}_k : L^m_\delta \to L^p_{-\delta}$$

is continuous, and

$$\|\mathcal{G}_k\|_{\mathcal{L}(L^m_\delta, L^p_{-\delta})} \leq C w(k)^{3(\frac{1}{m}-\frac{1}{p})-1}.$$

Proof. By the mean value theorem,

$$\begin{aligned}|(\mathcal{G}_{k+h}f - \mathcal{G}_k f)(x)| &= \left| \int_{\mathbb{R}^3} \frac{e^{i(k+h)|x-y|} - e^{ik|x-y|}}{4\pi|x-y|} f(y)\,dy \right| \\ &= \left| \int_{\mathbb{R}^3} \frac{h}{4\pi} \frac{de^{it}}{dt}\bigg|_{t=\xi} f(y) w(y)^\delta w(y)^{-\delta}\,dy \right| \leq \frac{|h|}{4\pi} \|f\|_{L^m_\delta} \|1\|_{L^{m'}_{-\delta}}\end{aligned}$$

for some $\xi = \xi(x,y,k,h)$ between $k|x-y|$ and $(k+h)|x-y|$. Thus,

$$\begin{aligned}\|\mathcal{G}_{k+h} - \mathcal{G}_k\|^p_{\mathcal{L}(L^m_\delta, L^p_{-\delta})} &= \sup_{\|f\|_{L^m_\delta} \leq 1} \int_{\mathbb{R}^3} |(\mathcal{G}_{k+h}f - \mathcal{G}_k f)(x)|^p w(x)^{-p\delta}\,dx \\ &\leq Ch^p \|1\|_{L^{m'}_{-\delta}} \|1\|_{L^p_{-\delta}} \xrightarrow{h \to 0} 0\end{aligned}$$

by (1.15) and the assumption about δ. Therefore, $\mathcal{G}_k$ is continuous with respect to k.

For $|k| > 1$, we use the estimate [Päi04, Thm. 3.1]

$$\|\mathcal{G}_k\|_{\mathcal{L}(L^m_\delta(\mathbb{R}^n), L^p_{-\delta}(\mathbb{R}^n))} \leq C|k|^{n(\frac{1}{m}-\frac{1}{p})-1},$$

now with $n = 3$. If the exponent is positive, this immediately proves the claim. If not, we combine this estimate with the fact that by continuity, $\mathcal{G}_k$ is bounded for k in the compact set $[-1,1]$. Thus by (1.16),

$$\|\mathcal{G}_k\|_{\mathcal{L}(L^m_\delta, L^p_{-\delta})} \leq C \min\{1, |k|^{3(\frac{1}{m}-\frac{1}{p})-1}\} \leq C w(k)^{3(\frac{1}{m}-\frac{1}{p})-1}.$$

□

For showing the invertibility of the Lippmann-Schwinger operator $I + \mathcal{G}_k Q$, we shall use weighted Sobolev spaces, for which we first demonstrate a few facts.

Definition 2.5. *For* $m \in \mathbb{N}$ *and* $\delta \in \mathbb{R}$, *the weighted Sobolev norm over a domain* $\Omega \subset \mathbb{R}^n$ *is defined by*

$$\|f\|_{W^{m,p}_\delta(\Omega)} = \left[\int_\Omega w(x)^{p\delta} \sum_{|\alpha| \leq m} |\partial^\alpha f(x)|^p\,dx \right]^{1/p} \qquad \text{when } 1 \leq p < \infty$$

$$\|f\|_{W^{m,\infty}_\delta(\Omega)} = \max_{|\alpha| \leq m} \|\partial^\alpha f\|_{L^\infty_\delta(\Omega)}.$$

The corresponding weighted Sobolev space consists of those functions whose norm is finite:

$$W^{m,p}_\delta(\Omega) = \{f \in \mathcal{S}'(\Omega) \mid \|f\|_{W^{m,p}(\Omega)} < \infty\}.$$

We also use the notation

$$W^{m,p}_{\mathrm{loc}} = \{f \in \mathcal{S}'(\mathbb{R}^3) \mid \|f\|_{W^{m,p}(K)} < \infty \textit{ for all compact subsets } K \subset \mathbb{R}^3\}.$$

In the case $p = 2$ we also write

$$H^m = W^{m,2} \qquad H^m_\delta = W^{m,2}_\delta \qquad H^m_{\mathrm{loc}} = W^{m,2}_{\mathrm{loc}}.$$

It might seem equally natural to define these norms the other way around, taking the derivatives only after multiplying by the weight. As is well known, this would not change the spaces.

Lemma 2.6. *For $1 \le p < \infty$, the weighted Sobolev norm is equivalent to the regular Sobolev norm of the weighted function:*

$$\|f\|_{W^{m,p}_\delta(\Omega)} \sim \|w^\delta f\|_{W^{m,p}(\Omega)}.$$

Proof. First show that $\|w^\delta f\|_{W^{m,p}(\Omega)} \le C\|f\|_{W^{m,p}_\delta(\Omega)}$: By the Leibniz formula and (1.16)

$$\begin{aligned}\|w^\delta f\|^p_{W^{m,p}(\Omega)} &\le \int_\Omega \sum_{|\alpha|\le m} \left[\sum_{\beta\le\alpha} \binom{\alpha}{\beta} \left|\partial^{\alpha-\beta} w(x)^\delta\right| \left|\partial^\beta f(x)\right|\right]^p dx \\ &\le C \int_\Omega w(x)^\delta \sum_{|\beta|\le m} \left|\partial^\beta f(x)\right|^p dx = C\|f\|^p_{W^{m,p}_\delta(\Omega)}.\end{aligned}$$

For the converse estimate, again use the Leibniz formula and Lemma 1.1 to get

$$\begin{aligned} w(x)^\delta|\partial^\alpha f(x)| &= \left|\partial^\alpha\left[w(x)^\delta f(x)\right] - \sum_{\alpha'<\alpha} \binom{\alpha}{\alpha'} \partial^{\alpha-\alpha'} w(x)^\delta \partial^{\alpha'} f(x)\right| \\ &\le \left|\partial^\alpha\left[w(x)^\delta f(x)\right]\right| + \sum_{\alpha'<\alpha} \binom{\alpha}{\alpha'} |\partial^{\alpha-\alpha'} w(x)^\delta||\partial^{\alpha'} f(x)| \\ &\le \left|\partial^\alpha\left[w(x)^\delta f(x)\right]\right| + C \sum_{\alpha'<\alpha} \binom{\alpha}{\alpha'} w(x)^\delta|\partial^{\alpha'} f(x)|.\end{aligned}$$

Note that now $|\alpha'| \le m-1$. Repeating the estimate m times gives the statement. □

Lemma 2.7. *The embedding $H^2_{\delta'}(\mathbb{R}^3) \hookrightarrow L^p_\delta(\mathbb{R}^3)$ is compact for all $p \in [2,6]$ and $\delta, \delta' \in \mathbb{R}$ with $\delta < \delta'$.*

Proof. Choose $\chi \in C_0^\infty(\mathbb{R})$ such that $\operatorname{supp}\chi \subset [-2,2]$ and $\chi \equiv 1$ on $[-1,1]$. Set $\chi_R(x) = \chi(|x| - R + 1)$ so that $\chi_R \in C_0^\infty(\mathbb{R}^3)$ and $\operatorname{supp}\chi_R \subset B(0,R+1)$ and $\chi_R|_{B(0,R)} \equiv 1$. Then the operator $\chi_R \cdot : H^2_{\delta'} \to H^2(B_R) \hookrightarrow L^p(B_R) \to L^p_\delta$ is compact by the Sobolev embedding theorem [Ada75, Thm. 6.2.II]. Here $\to$ denotes a continuous mapping and $\hookrightarrow$ a compact embedding.

It suffices to show that the operator $1 - \chi_R : H^2_{\delta'} \to L^p_\delta$ tends to zero as $R \to \infty$: by another Sobolev embedding theorem [Ada75, Thm. 5.4.I] and Lemma 1.1,

$$\begin{aligned}
\|(1-\chi_R)f\|_{L^p_\delta} &\le C\|w^\delta(1-\chi_R)f\|_{H^1} \\
&\le C\Big[\|w^\delta(1-\chi_R)f\|_{L^2} + \|(\nabla w^\delta)(1-\chi_R)f\|_{L^2} \\
&\qquad + \|w^\delta[\nabla(1-\chi_R)]f\|_{L^2} + \|w^\delta(1-\chi_R)\nabla f\|_{L^2}\Big] \\
&\le C\Big[\|w^\delta f\|_{L^2(\mathbb{R}^3\setminus B(0,R))} + \|w^\delta \nabla f\|_{L^2(\mathbb{R}^3\setminus B(0,R))}\Big] \\
&\le C w(R)^{\delta-\delta'}\|f\|_{H^2_{\delta'}},
\end{aligned}$$

where C does not depend on R. Therefore $\|1-\chi_R\|_{\mathcal{L}(H^2_{\delta'}, L^p_\delta)} \le C w(R)^{\delta-\delta'} \to 0$. □

For proving the invertibility of $I + \mathcal{G}_k Q$, we shall express the convolution operator $\mathcal{G}_k = \Phi(\cdot,k)*$ as multiplication on the Fourier transform side. To this end, we review the calculation of its Fourier transform with respect to $x \in \mathbb{R}^3$:

Lemma 2.8. *For $z \in \mathbb{C}_+$ and $k \in \mathbb{R}$,*

$$\begin{aligned}
F\Phi(\xi,z) &= \frac{1}{|\xi|^2 - z^2} \\
F\Phi(\xi,k) &= \lim_{\varepsilon \searrow 0} \frac{1}{|\xi|^2 - (k+i\varepsilon)^2} =: \frac{1}{|\xi|^2 - (k+i0)^2} \\
F\Phi(\xi,k) &= \lim_{\varepsilon \searrow 0} \frac{1}{|\xi|^2 - k^2 - i\varepsilon \operatorname{sgn} k} =: \frac{1}{|\xi|^2 - k^2 - i0 \operatorname{sgn} k}.
\end{aligned} \tag{2.3}$$

In the case $k = 0$, the limit reduces to $F\Phi(\xi,0) = |\xi|^{-2}$.

Proof. Integrating by parts in polar coordinates and substituting $t = \cos\theta$, we get

$$F\left[\frac{e^{iz|x|}}{4\pi|x|}\right](\xi) = \int_{S^2}\int_0^\infty \frac{\exp[i(z-\hat{x}\cdot\xi)r]}{4\pi r} r^2\,dr\,dS(\hat{x})$$
$$= \int_{S^2} -\frac{1}{4\pi(z-\hat{x}\cdot\xi)^2}\,dS(\hat{x})$$
$$= -\int_0^\pi\int_0^{2\pi} d\varphi\,\frac{1}{4\pi(z-|\xi|\cos\theta)^2}\sin\theta\,d\theta$$
$$= \frac{1}{2}\int_1^{-1}\frac{1}{(z-|\xi|t)^2}\,dt$$
$$= \frac{1}{|\xi|^2-z^2}.$$

Since $e^{-\varepsilon|x|}\varphi(x) \to \varphi(x)$ in $C(\mathbb{R}^3)$ as $\varepsilon \searrow 0$, and Φ is a tempered distribution of order zero,

$$\langle\Phi,\varphi\rangle = \lim_{\varepsilon\searrow 0}\left\langle\Phi, e^{-\varepsilon|\cdot|}\varphi\right\rangle = \lim_{\varepsilon\searrow 0}\left\langle\frac{e^{i(k+i\varepsilon)|x|}}{4\pi|x|},\varphi(x)\right\rangle_x,$$

i.e., $\Phi(x,k)$ is the distribution limit as $\varepsilon\searrow 0$ of the integrable functions $\Phi(x,k+i\varepsilon)$. Thus,

$$F\Phi(\xi) = \lim_{\varepsilon\searrow 0} F\left[\frac{e^{i(k+i\varepsilon)|x|}}{4\pi|x|}\right](\xi) = \lim_{\varepsilon\searrow 0}\frac{1}{|\xi|^2-(k+i\varepsilon)^2}.$$

The alternate form for $F\Phi(\xi,k)$ follows analogously since, if we take the square root with positive imaginary part,

$$\langle\Phi,\varphi\rangle = \lim_{\varepsilon\searrow 0}\left\langle\Phi, e^{i[(k^2+i\varepsilon\,\mathrm{sgn}\,k)^{1/2}-k]|\cdot|}\varphi\right\rangle = \lim_{\varepsilon\searrow 0}\left\langle\frac{e^{i(k^2+i\varepsilon\,\mathrm{sgn}\,k)^{1/2}|x|}}{4\pi|x|},\varphi(x)\right\rangle_x.$$

□

Lemma 2.9. *Let $p\in[2,6]$, $\delta>3/2$, $q\in L^\infty_\gamma$ for some $\gamma>2\delta+3/2$, and assume that q is such that the operator $I+\mathcal{G}_kQ$ is injective on $L^2_{-\delta}$. Then the operator $I+\mathcal{G}_kQ$ is invertible in $L^p_{-\delta}$, and in particular, the frequency domain Lippmann-Schwinger equation has a unique solution.*

Proof. We first note that $I+\mathcal{G}_kQ$ is also injective on $L^p_{-\delta}$: If $(I+\mathcal{G}_kQ)\psi = 0$ with $\psi\in L^p_{-\delta}$, Lemmata 2.3 and 2.4 imply that $\psi = -\mathcal{G}_kQ\psi\in L^2_{-\delta}$ and thus by assumption, $\psi = 0$.

By the Fredholm Alternative it suffices to show that $\mathcal{G}_kq$ is a compact operator on $L^p_{-\delta}$, since then injectivity will imply invertibility for the resolvent. This will be done by fixing any $\delta'\in(3/2,\min\{\delta,(\gamma-1)/2\})$ and showing that

$$L^p_{-\delta}\xrightarrow{Q}L^2_{\delta'}\xrightarrow{\mathcal{G}_k}H^2_{-\delta'}\hookrightarrow L^p_{-\delta},$$

where $\longrightarrow$ denotes a continuous mapping and $\hookrightarrow$ a compact embedding.

The continuity of $Q : L^p_{-\delta} \to L^2_{\delta'}$ is asserted by Lemma 2.3. The compactness of the embedding $H^2_{-\delta'} \hookrightarrow L^p_{-\delta}$ was proved in Lemma 2.7.

For proving the continuity of $\mathcal{G}_k : L^2_{\delta'} \to H^2_{-\delta'}$, use Lemma 2.6 to see that

$$\begin{aligned}
\|\mathcal{G}_k f\|_{H^2_{-\delta'}} &\le C\|w^{-\delta'}\mathcal{G}_k f\|_{H^2} \\
&\le C\|w^2 F(w^{-\delta'}\mathcal{G}_k f)\|_{L^2} \\
&\le C\|F(w^{-\delta'}\mathcal{G}_k f)\|_{L^2} + C\||\cdot|^2 F(w^{-\delta'}\mathcal{G}_k f)\|_{L^2} \\
&= C\|\mathcal{G}_k f\|_{L^2_{-\delta'}} + C\|\triangle(w^{-\delta'}\mathcal{G}_k f)\|_{L^2}.
\end{aligned}$$

The first term is estimated using Lemma 2.4. As for the second term,

$$\begin{aligned}
\|\triangle(w^{-\delta'}\mathcal{G}_k f)\|_{L^2} &\le C\left[\|(\triangle w^{-\delta'})\mathcal{G}_k f\|_{L^2} + \|(\nabla w^{-\delta'})\cdot\nabla\mathcal{G}_k f\|_{L^2} + \|w^{-\delta'}\triangle\mathcal{G}_k f\|_{L^2}\right] \\
&\le C\left[\|\mathcal{G}_k f\|_{L^2_{-\delta'}} + \sum_{j=1}^{3}\|\partial_j\mathcal{G}_k f\|_{L^2_{-\delta'}} + \|\triangle\mathcal{G}_k f\|_{L^2_{-\delta'}}\right]
\end{aligned} \tag{2.4}$$

by Lemma 1.1. For estimating the first order terms, choose $\chi \in C_0^\infty(B_{3|k|+3}, [0,1])$ such that $\chi \equiv 1$ in $B_{2|k|+2}$, and write $\partial_j\mathcal{G}_k f = g_1 + g_2$, where

$$(Fg_1)(\xi) = \frac{(Ff)(\xi)}{|\xi|^2 - (k+i0)^2} i\xi_j\chi(\xi), \quad (Fg_2)(\xi) = \frac{(Ff)(\xi)}{|\xi|^2 - (k+i0)^2} i\xi_j[1-\chi(\xi)]. \tag{2.5}$$

Now the fact that multiplication by $\xi\chi(\xi) \in C_0^\infty$ is a bounded operator on $H^{-\delta'}$ gives

$$\|g_1\|_{L^2_{-\delta'}} = \|Fg_1\|_{H^{-\delta'}} \le C\|F(\mathcal{G}_k f)\|_{H^{-\delta'}} = C\|\mathcal{G}_k f\|_{L^2_{-\delta'}}.$$

Also,

$$\varphi(\xi) := \frac{\xi_j}{|\xi|^2 - (k+i0)^2}[1-\chi(\xi)]$$

is a bounded function, and thus

$$\|g_2\|_{L^2_{-\delta'}} \le C\|g_2\|_{L^2} = C\|Fg_2\|_{L^2} \le C\|\varphi\|_{L^\infty}\|Ff\|_{L^2} = C\|f\|_{L^2} \le C\|f\|_{L^2_{\delta'}}.$$

Therefore Lemma 2.4 shows that

$$\|\partial_j\mathcal{G}_k f\|_{L^2_{-\delta'}} \le C\left[\|g_1\|_{L^2_{-\delta'}} + \|g_2\|_{L^2_{-\delta'}}\right] \le C\|f\|_{L^2_{\delta'}}.$$

The Laplacian term in (2.4) is estimated exactly in the same way, except with multiplication by $|\xi|^2$ instead of ξ_j in (2.5). □

The injectivity condition is automatically satisfied for nonzero frequencies k, as is shown by the following lemma. For the zero frequency, we need to explicitly assume injectivity.

Lemma 2.10. *Let $\delta > 3/2$ and $q \in L^\infty_\gamma(\mathbb{R}^3, \mathbb{R})$ for some $\gamma > \max\{\delta + 7/2, 2\delta + 1\}$. Then the operator $I + \mathcal{G}_k Q$ is injective on $L^2_{-\delta}$ for all $k \neq 0$.*

Proof. Fix $k \neq 0$ and assume that $L^2_{-\delta} \ni \varphi = -\mathcal{G}_k q\varphi$. Thus

$$(-\triangle - k^2 + q)\varphi = (-\triangle - k^2)(-\mathcal{G}_k q\varphi) + q\varphi = -q\varphi + q\varphi = 0.$$

Since the operator $-\triangle - k^2 + q$ is injective in L^2 when $q(x) = \mathcal{O}(|x|^{-\rho})$, $\rho > 1$ [Kat59], it suffices to show that $\varphi \in L^2$.

Take a sequence of functions $\psi_n \in C_0^\infty$ converging to $q\varphi$ in L^2_δ. Then

$$0 = \operatorname{Im}(\varphi, q\varphi) = \operatorname{Im}(\mathcal{G}_k q\varphi, q\varphi) = \lim_{n\to\infty} \operatorname{Im}(\mathcal{G}_k \psi_n, \psi_n) = \lim_{n\to\infty} \operatorname{Im} \int_{\mathbb{R}^3} \frac{|F\psi_n(\xi)|^2}{|\xi|^2 - k^2 \mp i0}\, d\xi,$$

where $\mp = -\operatorname{sgn} k$. Split the integral into two parts using a smooth cut-off function $\chi \in C_0^\infty(B_{|k|/2}, [0,\infty))$ with $\chi \equiv 1$ in $B_{|k|/3}$: The part around the origin is away from the singularity, and thus

$$\operatorname{Im} \int_{\mathbb{R}^3} \frac{\chi(\xi)|F\psi_n(\xi)|^2}{|\xi|^2 - k^2 \mp i0}\, d\xi = \operatorname{Im} \int_{\mathbb{R}^3} \frac{\chi(\xi)|F\psi_n(\xi)|^2}{|\xi|^2 - k^2}\, d\xi = 0.$$

Write the singular part using pullback with the smooth $f(\xi) = |\xi|^2 - k^2$, whose gradient does not vanish for $|\xi| > |k|/3$:

$$\frac{1}{|\xi|^2 - k^2 \mp i0} = (f^* g)(\xi), \qquad g(\xi) = \left(\frac{1}{\xi \mp i0}\right) = \operatorname{pv}\tfrac{1}{\xi} \pm i\pi\delta_0(\xi).$$

Therefore

$$\begin{aligned} 0 &= \lim_{n\to\infty} \operatorname{Im}\left[\lim_{\varepsilon\searrow 0} \int_{||\xi|^2 - k^2| > \varepsilon} \frac{1-\chi(\xi)}{|\xi|^2 - k^2} |F\psi_n(\xi)|^2\, d\xi \pm i\pi \langle f^*\delta_0 (1-\chi) F\psi_n, F\psi_n\rangle\right] \\ &= \pm\pi \lim_{n\to\infty} \int_{|\xi|^2 = k^2} |F\psi_n(\xi)|^2\, dS(\xi) \qquad (2.6) \\ &= \pm\pi \int_{|\xi| = |k|} |F(q\varphi)(\xi)|^2\, dS(\xi) \end{aligned}$$

by the Sobolev trace theorem, since $F\psi_n \to F(q\varphi)$ in H^δ.

We shall now use (2.6) to analyse the behaviour of $F(q\varphi)$ near the sphere $|\xi| = |k|$, where $F\Phi$ is singular: Let $\alpha > 5/2$. Let $\mathcal{K}_s : L^2(\mathbb{R}^3) \to L^2(S^2)$ be the operator defined by

$$\begin{aligned} [\mathcal{K}_s(w^\alpha q\varphi)](\omega) &= F(q\varphi)(s\omega) \\ &= F(q\varphi)(s\omega) - F(q\varphi)(|k|\omega) \\ &= \int_{\mathbb{R}^3} K_s(x,\omega) w(x)^\alpha q(x)\varphi(x)\, dx, \end{aligned}$$

where
$$K_s(x,\omega) = e^{is\omega\cdot x}\left[e^{i(|k|-s)\omega\cdot x} - 1\right] w(x)^{-\alpha}.$$

We can estimate the operator norm of $\mathcal{K}_s$ through its Hilbert-Schmidt norm as follows: because the chord line of the unit circle is not longer than the corresponding arc, we have $|e^{i(|k|-s)\omega\cdot x} - 1| \leq |(|k|-s)\omega\cdot x| \leq ||k|-s||x|$, and thus

$$\begin{aligned}
\|\mathcal{K}_s\|^2_{\mathcal{L}(L^2(\mathbb{R}^3),L^2(S^2))} &\leq \int_{S^2}\int_{\mathbb{R}^3} \left|e^{-is\omega\cdot x}\right|^2 \left|e^{i(|k|-s)\omega\cdot x} - 1\right|^2 w(x)^{-2\alpha}\,dx\,dS(\omega) \\
&\leq (4\pi)^2 \left||k|-s\right|^2 \int_0^\infty r^2 w(r)^{-2\alpha} r^2\,dr \\
&\leq C\left||k|-s\right|^2.
\end{aligned}$$

If we choose $\alpha \in (5/2, \gamma-\delta)$, we see that $\|w^\alpha q\varphi\|_{L^2} \leq C\|\varphi\|_{L^2_{\alpha-\gamma}} \leq C\|\varphi\|_{L^2_{-\delta}} < \infty$ and thus

$$\|[F(q\varphi)](s\cdot)\|_{L^2(S^2)} \leq C\left||k|-s\right|, \tag{2.7}$$

the constant now containing the norm of φ.

Write

$$\|\varphi\|^2_{L^2} = \|\mathcal{G}_k q\varphi\|^2_{L^2} = C\|F(\mathcal{G}_k q\varphi)\|^2_{L^2} = C\int_{\mathbb{R}^3}\left|\frac{F(q\varphi)(\xi)}{|\xi|^2-(k+i0)^2}\right|^2 d\xi.$$

Split the domain into $B_{2|k|}$ and its complement. In $B_{2|k|}$, use (2.7) and further estimate $||k|-r|^2 \leq ||k|-r+i\varepsilon|^2 = (r-|k|-i\varepsilon)(r-|k|+i\varepsilon)$ to see that

$$\begin{aligned}
\int_{B_{2|k|}}\left|\frac{F(q\varphi)(\xi)}{|\xi|^2-(k+i0)^2}\right|^2 d\xi &= \lim_{\varepsilon\searrow 0}\int_0^{2|k|}\frac{1}{|r^2-(k+i\varepsilon)^2|^2}\int_{S^2}|F(q\varphi)(r\omega)|^2\,dS(\omega)\,r^2\,dr \\
&\leq \lim_{\varepsilon\searrow 0} C\int_0^{2|k|}\frac{(r-|k|-i\varepsilon)(r-|k|+i\varepsilon)\,r^2\,dr}{(r+k+i\varepsilon)(r+k-i\varepsilon)(r-k-i\varepsilon)(r-k+i\varepsilon)} \\
&= \lim_{\varepsilon\searrow 0} C\int_0^{2|k|}\frac{r^2\,dr}{(r+|k|+i\varepsilon)(r+|k|-i\varepsilon)} \\
&= C\int_0^{2|k|}\frac{r^2}{(r+|k|)^2}\,dr \\
&< \infty.
\end{aligned}$$

In $\mathbb{R}^3\setminus B_{2|k|}$, simply estimate $|\xi|^2-(k+i\varepsilon)^2 \geq k^2$ to get

$$\int_{\mathbb{R}^3\setminus B_{2|k|}}\left|\frac{F(q\varphi)(\xi)}{|\xi|^2-(k+i0)^2}\right|^2 d\xi \leq \frac{\|F(q\varphi)\|^2_{L^2}}{k^2} = \frac{C\|q\varphi\|^2_{L^2}}{k^2} \leq \frac{C\|\varphi\|^2_{L^2_{-\delta}}}{k^2} < \infty.$$

□

Lemma 2.11. *Let $2 \le p < 6$, $\delta > 3/2$, $q \in L^\infty_\gamma(\mathbb{R}^3,\mathbb{R})$ for some $\gamma > \max\{\delta+7/2, 2\delta+1\}$, and assume that q is such that the operator $I+\mathcal{G}_0 Q$ is injective on $L^2_{-\delta}$. Then the operator valued function*

$$\mathbb{R} \ni k \mapsto (I+\mathcal{G}_k Q)^{-1} \in \mathcal{L}(L^p_{-\delta})$$

is continuous and bounded with respect to k.

Proof. By Lemmata 2.3 and 2.4,

$$\|\mathcal{G}_k Q\|_{\mathcal{L}(L^p_{-\delta},L^p_{-\delta})} \le \|\mathcal{G}_k\|_{\mathcal{L}(L^m_\delta,L^p_{-\delta})}\|Q\|_{\mathcal{L}(L^p_{-\delta},L^m_\delta)} \le Cw(k)^{3(\frac{1}{m}-\frac{1}{p})-1} \to 0$$

as $|k| \to \infty$, if we choose $m \in (3p/(3+p), 2]$, which is possible by our assumption about p. Thus, when $|k|$ is sufficiently large, say $|k| \ge M$, this norm is less than 1, and the resolvent is given by the Neumann series

$$(I+\mathcal{G}_k Q)^{-1} = \sum_{j=0}^{\infty} (-\mathcal{G}_k Q)^j$$

for which

$$\|\, (I+\mathcal{G}_k Q)^{-1} \,\|_{\mathcal{L}(L^p_{-\delta})} \le \frac{1}{1-CM^{1-3(\frac{1}{m}-\frac{1}{p})}}.$$

Continuity follows, since

$$\begin{aligned} \|(I+\mathcal{G}_k Q)^{-1} - (I+\mathcal{G}_{k_0} Q)^{-1}\|_{\mathcal{L}(L^p_{-\delta})} &= \left\|(I+\mathcal{G}_k Q)^{-1}\left[\mathcal{G}_{k_0}-\mathcal{G}_k\right] Q(I+\mathcal{G}_{k_0}Q)^{-1}\right\| \\ &\le C\|\mathcal{G}_{k_0}-\mathcal{G}_k\|_{\mathcal{L}(L^m_\delta,L^p_{-\delta})} \to 0 \end{aligned} \tag{2.8}$$

as $k \to k_0$, by Lemma 2.4.

Then consider the case $|k| \le M$. We first show local boundedness using the formula

$$\|(A+b)^{-1}\| = \|A^{-1}(I+bA^{-1})^{-1}\| \le \frac{\|A^{-1}\|}{1-\|b\|\,\|A^{-1}\|}:$$

Fix $k_0 \in [-M,M]$ and set $A = I+\mathcal{G}_{k_0}Q$ and $b = \mathcal{G}_k Q - \mathcal{G}_{k_0}Q$. Then by Lemmata 2.9 and 2.10 and our assumption, A is invertible. Thus by Lemma 2.4,

$$\|b\|_{\mathcal{L}(L^p_{-\delta})} \le \|\mathcal{G}_k - \mathcal{G}_{k_0}\|_{\mathcal{L}(L^m_\delta,L^p_{-\delta})}\|Q\|_{\mathcal{L}(L^p_{-\delta},L^m_\delta)} < \frac{1}{2\,\|(I+\mathcal{G}_{k_0}Q)^{-1}\|_{\mathcal{L}(L^p_{-\delta})}}$$

for k in some small neighbourhood of k_0. For such k, we thus have

$$\|(I+\mathcal{G}_k Q)^{-1}\|_{\mathcal{L}(L^p_{-\delta})} \le \frac{\|(I+\mathcal{G}_{k_0}Q)^{-1}\|_{\mathcal{L}(L^p_{-\delta})}}{1-1/2},$$

and (2.8) shows continuity. Boundedness follows, since the compact set $[-M,M]$ can be covered by a finite number of such neighbourhoods. □

Remark. Combining Lemmata 2.4 and 2.11 shows that the resolvent of the Schrödinger operator

$$\mathcal{G}_k^q := \lim_{\varepsilon \searrow 0} \left(-\triangle + Q - (k+i\varepsilon)^2\right)^{-1} = (I+\mathcal{G}_k Q)^{-1}\mathcal{G}_k$$

satisfies the same estimates as $\mathcal{G}_k$ in Lemma 2.4.

For estimating the derivatives

$$\begin{aligned}
\partial_k \hat{u}_s &= -\left[\partial_k (I+\mathcal{G}_k Q)^{-1}\right]\mathcal{G}_k(q\hat{u}) - (I+\mathcal{G}_k Q)^{-1}\left[\partial_k \mathcal{G}_k\right](q\hat{u}) - (I+\mathcal{G}_k Q)^{-1}\mathcal{G}_k(q\partial_k \hat{u})\\
\partial_k^2 \hat{u}_s &= -\left[\partial_k^2 (I+\mathcal{G}_k Q)^{-1}\right]\mathcal{G}_k(q\hat{u}) - (I+\mathcal{G}_k Q)^{-1}\left[\partial_k^2 \mathcal{G}_k\right](q\hat{u}) - (I+\mathcal{G}_k Q)^{-1}\mathcal{G}_k(q\partial_k^2 \hat{u})\\
&\quad - 2\left[\partial_k (I+\mathcal{G}_k Q)^{-1}\right]\left[\partial_k \mathcal{G}_k\right](q\hat{u}) - 2\left[\partial_k (I+\mathcal{G}_k Q)^{-1}\right]\mathcal{G}_k(q\partial_k \hat{u})\\
&\quad - 2(I+\mathcal{G}_k Q)^{-1}\left[\partial_k \mathcal{G}_k\right](q\partial_k \hat{u})
\end{aligned}$$

we shall use the following estimates for the k derivatives of the operators $\mathcal{G}_k$ and $(I+\mathcal{G}_k Q)^{-1}$.

Lemma 2.12. *The operator valued function*

$$\mathbb{R} \ni k \mapsto \mathcal{G}_k : L^2_\delta \to L^p_{-\delta}$$

has two bounded derivatives for $2 \le p < \infty$ *and* $\delta > 5/2$.

Proof. Show that we can differentiate under the convolution integral: $\partial_k \mathcal{G}_k = \partial_k(\Phi *) = (\partial_k \Phi)*$, *i.e.*, that $(\partial_k \Phi)*$ is the operator norm limit of the difference quotient. By the mean value theorem, there is a $\xi = \xi(x,y,k,h)$ between k and $k+h$ such that

$$\begin{aligned}
&\left\| \left(\frac{\mathcal{G}_{k+h} - \mathcal{G}_k}{h} - (\partial_k \Phi)* \right) f \right\|_{L^p_{-\delta}}^p \\
&\quad = \int_{\mathbb{R}^3} \left| \int_{\mathbb{R}^3} \left[\frac{e^{i(k+h)|x-y|} - e^{ik|x-y|}}{4\pi h|x-y|} - \frac{ie^{ik|x-y|}}{4\pi} \right] f(y)\,dy \right|^p w(x)^{-p\delta} dx \\
&\quad = \int_{\mathbb{R}^3} \left| \int_{\mathbb{R}^3} \frac{ie^{ik|x-y|}}{4\pi} \left[e^{i(\xi-k)|x-y|} - 1 \right] f(y)\,dy \right|^p w(x)^{-p\delta} dx \\
&\quad \le C \int_{\mathbb{R}^3} \left\| \frac{e^{i(\xi-k)|x-\cdot|} - 1}{w(x)^\varepsilon w^\varepsilon} f w^\delta \right\|_{L^2}^p \|w^{\varepsilon-\delta}\|_{L^2}^p w(x)^{p(\varepsilon-\delta)} dx.
\end{aligned}$$

Now $|e^{i(\xi-k)|x-y|} - 1| \le |\xi - k||x-y| \le h(|x|+|y|)$, and of course also ≤ 2. Thus if $\varepsilon \ge 0$,

$$\left| \frac{e^{i(\xi-k)|x-y|} - 1}{w(x)^\varepsilon w(y)^\varepsilon} \right| \le \begin{cases} \frac{2}{(1+\frac{1}{h})^{\varepsilon/2}} = \frac{2h^{\varepsilon/2}}{(1+h)^{\varepsilon/2}} \le 2h^{\varepsilon/2} & \text{when } |x| > \frac{1}{\sqrt{h}} \text{ or } |y| > \frac{1}{\sqrt{h}} \\ \frac{h(|x|+|y|)}{w(x)^\varepsilon w(y)^\varepsilon} \le h\frac{2}{\sqrt{h}} \le 2\sqrt{h} & \text{when } |x|,|y| \le \frac{1}{\sqrt{h}} \end{cases} \tag{2.9}$$

and consequently by (1.15),

$$\left\| \left(\frac{\mathcal{G}_{k+h} - \mathcal{G}_k}{h} - \partial_k \mathcal{G}_k \right) f \right\|_{L^p_{-\delta}} \leq C(h^{\varepsilon/2} + h^{1/2}) \|f\|_{L^2_\delta} \|w^{\varepsilon-\delta}\|_{L^2} \|w^{\varepsilon-\delta}\|_{L^p}$$
$$\leq C(h^{\varepsilon/2} + h^{1/2}) \|f\|_{L^2_\delta},$$

which proves the claim about differentiating under the integral if we choose $\varepsilon \in (0, \delta - 3/2)$.

Then do the same for the second derivative $\partial_k^2 \Phi(x-y,k) = -e^{ik|x-y|}|x-y|/4\pi$:

$$\left\| \left(\frac{\partial_k \mathcal{G}_{k+h} - \partial_k \mathcal{G}_k}{h} - (\partial_k^2 \Phi)* \right) f \right\|^p_{L^p_{-\delta}}$$
$$= \int_{\mathbb{R}^3} \left| \int_{\mathbb{R}^3} \left[-\frac{e^{i(k+h)|x-y|} - e^{ik|x-y|}}{4\pi h} + \frac{e^{ik|x-y|}}{4\pi} |x-y| \right] f(y)\, dy \right|^p w(x)^{-p\delta} dx$$
$$= \int_{\mathbb{R}^3} \left| \int_{\mathbb{R}^3} \frac{e^{i\xi|x-y|}}{4\pi} \left[e^{i(k-\xi)|x-y|} - 1 \right] |x-y| f(y)\, dy \right|^p w(x)^{-p\delta} dx$$
$$\leq C \int_{\mathbb{R}^3} \left\| \frac{e^{i(k-\xi)|x-\cdot|} - 1}{w(x)^\varepsilon w^\varepsilon} |x-y| f w^\delta \right\|^p_{L^2} \|w^{\varepsilon-\delta}\|^p_{L^2} w(x)^{p(\varepsilon-\delta)} dx$$

for some $\xi = \xi(x,y,k,h)$ between k and $k+h$. This time estimate as in (2.9), but now with $|x-y| \leq |x| + |y| \leq 2w(x)w(y)$:

$$\left| \frac{e^{i(k-\xi)|x-y|} - 1}{w(x)^\varepsilon w(y)^\varepsilon} |x-y| \right| \leq \left| 2 \frac{e^{i(k-\xi)|x-y|} - 1}{w(x)^{\varepsilon-1} w(y)^{\varepsilon-1}} \right| \leq 4 \left(h^{\frac{\varepsilon-1}{2}} + h^{\frac{1}{2}} \right),$$

and consequently

$$\left\| \left(\frac{\partial_k \mathcal{G}_{k+h} - \partial_k \mathcal{G}_k}{h} - (\partial_k^2 \Phi)* \right) f \right\|_{L^p_{-\delta}} \leq C \left(h^{\frac{\varepsilon-1}{2}} + h^{1/2} \right) \|w^{\varepsilon-\delta}\|_{L^2} \|w^{\varepsilon-\delta}\|_{L^p} \|f\|_{L^2_\delta}.$$

Now we have to choose $\varepsilon \in (1, \delta - 3/2)$.

The norms are bounded since

$$\|\partial_k \mathcal{G}_k f\|^p_{L^p_{-\delta}} \leq \int_{\mathbb{R}^3} \left(\int_{\mathbb{R}^3} \left| \frac{ie^{ik|x-y|}}{4\pi} \right| |f(y)|\, dy \right)^p w(x)^{-p\delta} dx \leq C \|1\|^p_{L^p_{-\delta}} \|1\|^p_{L^2_{-\delta}} \|f\|^p_{L^2_\delta}$$

and

$$\|\partial_k^2 \mathcal{G}_k f\|^p_{L^p_{-\delta}} \leq \int_{\mathbb{R}^3} \left(\int_{\mathbb{R}^3} \left| \frac{e^{ik|x-y|}}{4\pi} \right| |x-y| |f(y)|\, dy \right)^p w(x)^{-p\delta} dx$$
$$\leq C \int_{\mathbb{R}^3} \left(\int_{\mathbb{R}^3} |f(y)| w(y)^\delta w(y)^{1-\delta} dy \right)^p w(x)^{p(1-\delta)} dx$$
$$\leq C \|1\|^p_{L^p_{1-\delta}} \|1\|^p_{L^2_{1-\delta}} \|f\|^p_{L^2_\delta}.$$

□

Corollary 2.13. *Let $2 \le p < 6$, $\delta > 5/2$ and $q \in L^\infty_\gamma(\mathbb{R}^3,\mathbb{R})$, $\gamma > 2\delta + 1$ such $I + \mathcal{G}_0 Q$ is injective on $L^2_{-\delta}$. Then the operator valued function*

$$\mathbb{R} \ni k \mapsto (I + \mathcal{G}_k Q)^{-1} \in \mathcal{L}(L^p_{-\delta})$$

has two bounded derivatives.

Proof. Use the formula $\partial_k A^{-1} = -A^{-1}(\partial_k A)A^{-1}$ and its consequence

$$\partial_k^2 A^{-1} = 2A^{-1}(\partial_k A)A^{-1}(\partial_k A)A^{-1} - A^{-1}(\partial_k^2 A)A^{-1}$$

to obtain

$$\begin{aligned}
\|\partial_k (I + \mathcal{G}_k Q)^{-1}\|_{\mathcal{L}(L^p_{-\delta})} &\le \|(I + \mathcal{G}_k Q)^{-1}\|^2_{\mathcal{L}(L^p_{-\delta})} \|\partial_k \mathcal{G}_k\|_{\mathcal{L}(L^2_\delta, L^p_{-\delta})} \|Q\|_{\mathcal{L}(L^p_{-\delta}, L^2_\delta)}, \\
\|\partial_k^2 (I + \mathcal{G}_k Q)^{-1}\|_{\mathcal{L}(L^p_{-\delta})} &\le 2\|(I + \mathcal{G}_k Q)^{-1}\|^3_{\mathcal{L}(L^p_{-\delta})} \|\partial_k \mathcal{G}_k\|_{\mathcal{L}(L^2_\delta, L^p_{-\delta})} \|Q\|_{\mathcal{L}(L^p_{-\delta}, L^2_\delta)} \\
&\quad + \|(I + \mathcal{G}_k Q)^{-1}\|^2_{\mathcal{L}(L^p_{-\delta})} \|\partial_k^2 \mathcal{G}_k\|_{\mathcal{L}(L^2_\delta, L^p_{-\delta})} \|Q\|_{\mathcal{L}(L^p_{-\delta}, L^2_\delta)}.
\end{aligned}$$

The right hand sides are bounded by Lemmata 2.11 and 2.12. □

The norm estimates for the incident and scattered waves are summarized in the following corollary.

Corollary 2.14. *Let $m \in \{0,1,2\}$, $2 \le p < 6$ and $\delta > m + 3/2$. Assume that $q \in L^\infty_\gamma(\mathbb{R}^3,\mathbb{R})$ with $\gamma > 2\delta + 1$ is such that $I + \mathcal{G}_0 Q$ is injective on $L^2_{-\delta}$. Then if*

$$\left\| \frac{\partial^m \hat{u}_{\mathrm{i}}}{\partial k^m} \right\|_{L^p_{-\delta}} \le \frac{C}{|k|^b}, \tag{2.10}$$

we also have

$$\left\| \frac{\partial^m \hat{u}_{\mathrm{s}}}{\partial k^m} \right\|_{L^p_{-\delta}} \le \frac{C'}{|k|^b}. \tag{2.11}$$

In particular, the estimates (2.10) and (2.11) hold if

$$u_{\mathrm{i}}(x,t) = \int_{S^2} h(x \cdot \omega - t, \omega)\, dS(\omega)$$

with $h \in C_0^b(\mathbb{R}, L^2(S^2))$.

2.2 Regularity of solutions

Lemma 2.15. *Assume that* $q \in W^{j,\infty}_{\mathrm{loc}}$ *and that* $\hat{u}(\cdot,k) \in L^2_{\mathrm{loc}}$ *solves the frequency domain plasma wave equation (1.8). Then* $\hat{u}(\cdot,k) \in H^{2+j}_{\mathrm{loc}} \subset C^j(\mathbb{R}^3)$.

Proof. Let K be any bounded set in $\mathbb{R}^n$. By assumption,

$$\|\triangle \hat{u}(\cdot,k)\|_{L^2(K)} = \|(q-k^2)\hat{u}(\cdot,k)\|_{L^2(K)} \leq \left(k^2 + \|q\|_{L^\infty(K)}\right)\|\hat{u}(\cdot,k)\|_{L^2(K)} < \infty.$$

Thus by standard elliptic regularity results [Eva98, Thm 6.3.1], $\hat{u}(\cdot,k) \in H^2_{\mathrm{loc}}$. The statement for $j > 0$ follows iteratively: Assume that $\hat{u}(\cdot,k) \in H^m_{\mathrm{loc}}$. Then operating on (1.8) by ∂^α with $|\alpha| = m$ and and using the same estimate as above, we find that

$$\begin{aligned}\|\triangle \partial^\alpha \hat{u}(\cdot,k)\|_{L^2(K)} &= \left\| \sum_{\beta \leq \alpha} \binom{\alpha}{\beta} (\partial^\beta q)\partial^{\alpha-\beta}\hat{u}(\cdot,k) - k^2 \hat{u}(\cdot,k) \right\|_{L^2(K)} \\ &\leq \left(k^2 + 2^m \|q\|_{W^{m,\infty}(K)}\right) \|\hat{u}(\cdot,k)\|_{H^m(K)} < \infty.\end{aligned}$$

Thus again by elliptic regularity, $\partial^\alpha \hat{u}(\cdot,k) \in H^2_{\mathrm{loc}}$. Therefore, $\hat{u}(\cdot,k) \in H^{m+2}_{\mathrm{loc}}$. The inclusion in the claim is the standard Sobolev embedding result [Ada75, Thm. 5.4.I]. □

We note that *Green's representation formula*

$$\begin{aligned} f(x) = \int_{\partial B_R} &\left[\frac{\partial f}{\partial n}(y)\Phi(x-y) - f(y)\frac{\partial \Phi}{\partial n(y)}(x-y)\right] dS(y) \\ &- \int_{B_R} [(\triangle + k^2)f(y)]\Phi(x-y)dy \end{aligned} \tag{2.12}$$

for $|x| < R$, valid in the L^2 sense when $f \in H^1_{\mathrm{loc}}$ and $\triangle f \in L^2_{\mathrm{loc}}$, also holds pointwise when $f \in H^2_{\mathrm{loc}}$. This can be seen by approximating f with smooth functions, for which (2.12) is shown, *e.g.*, in [CK98, Thm. 2.1], and using the Sobolev embedding theorem [Ada75, Thm. 5.4.I].

2.3 Wave operator

A common formulation [LP67, RS79] for time domain scattering problems is in terms of the *wave operator* [RS79]

$$\Omega^\pm = \lim_{t \to \mp\infty} W_1(-t)W_0(t),$$

where W_0 and W_1 are the *propagators*

$$W_0(t) : \begin{pmatrix} u_i(\cdot,s) \\ \partial_t u_i(\cdot,s) \end{pmatrix} \mapsto \begin{pmatrix} u_i(\cdot,s+t) \\ \partial_t u_i(\cdot,s+t) \end{pmatrix} \qquad W_1(t) : \begin{pmatrix} u(\cdot,s) \\ \partial_t u(\cdot,s) \end{pmatrix} \mapsto \begin{pmatrix} u(\cdot,s+t) \\ \partial_t u(\cdot,s+t) \end{pmatrix}.$$

This definition is independent of the choice of $s \in \mathbb{R}$, since the free space wave equation and the plasma wave equation are time-independent.

Often one also considers the *scattering operator* $S = (\Omega^-)^*\Omega^+$ which relates to each other the asymptotical free space waves that the solution approaches before and after the scattering. We, however, shall not use it since we only deal with the incident free space wave and the scattered full wave $(u, \partial_t u) = \Omega^+(u_\mathrm{i}, \partial_t u_\mathrm{i})$. We shall show shortly in Lemma 2.19 that under certain assumptions, this formulation is equivalent to the time domain integral equation formulation (1.7).

A natural environment in which to work with the propagators is given in terms of the energies with and without potential: the propagators W_0 and W_1 preserve these energies, respectively. These energies and the associated Banach spaces are defined as follows:

Definition 2.16. *Denote by X the completion of $C_0^\infty(\mathbb{R}^3)$ with respect to the norm*

$$\|f_1\|_X := \|\nabla f_1\|_{L^2},$$

with inner product

$$(f_1, g_1)_X := (\nabla f_1, \nabla g_1)_{L^2}.$$

Set $H := X \times L^2$,

$$\left(\begin{pmatrix} f_1 \\ f_2 \end{pmatrix}, \begin{pmatrix} g_1 \\ g_2 \end{pmatrix} \right)_H := (f_1, g_1)_X + (f_2, g_2)_{L^2}, \qquad \left\| \begin{pmatrix} f_1 \\ f_2 \end{pmatrix} \right\|_H := \left(\begin{pmatrix} f_1 \\ f_2 \end{pmatrix}, \begin{pmatrix} f_1 \\ f_2 \end{pmatrix} \right)_H^{1/2}$$

and write

$$\tilde{H} := \left\{ u \in H^1(\mathbb{R}, L^2_{\mathrm{loc}}(\mathbb{R}^3)) \,\middle|\, \begin{pmatrix} u(\cdot,t) \\ \partial_t u(\cdot,t) \end{pmatrix} \in H \text{ for all } t \in \mathbb{R} \right\}.$$

For any $p \in L^\infty(\mathbb{R}^3, \mathbb{R})$, define on H the energy form

$$E_p(f_1, f_2) := \int_{\mathbb{R}^3} \left[|\nabla f_1(x)|^2 + |f_2(x)|^2 + p(x)|f_1(x)|^2 \right] dx$$

and on $\tilde{H}$ the energy of the wave at time t, $\tilde{E}_p(u,t) := E_p\big(u(\cdot,t), \partial_t u(\cdot,t)\big)$.

Note that the first components of elements of H are locally square integrable by the Gagliardo-Nirenberg-Sobolev inequality (see, *e.g.*, [Eva98, Thm. 5.6.1] or [Ada75, Sec. 5.11])

$$\|f\|_{L^{p^*}(\mathbb{R}^n)} \le C\|\nabla f\|_{L^p(\mathbb{R}^n)}, \qquad p < p^* = \frac{np}{n-p} < \infty. \tag{2.13}$$

With $n = 3$ and $p = 2$, $p^* = 6$ and we get for any bounded $V \subset \mathbb{R}^3$

$$\begin{aligned} \|f_1\|_{L^2(V)}^2 &= \|f_1^2\|_{L^1(V)} \\ &\le m(V)^{2/3}\|f_1^2\|_{L^3} \\ &\le C\|f_1\|_{L^6}^2 \\ &\le C\|\nabla f_1\|_{L^2}^2 \\ &\le C\|(f_1, f_2)\|_H^2. \end{aligned} \tag{2.14}$$

An essential fact that we shall use when showing the existence of the wave operator is the equivalence of the energy forms with and without potential. For this, we shall assume that the potential is such that the Schrödinger operator $-\triangle+Q$ does not have eigenvalues. A sufficient condition for this is that the potential only takes positive values; then it is easy to see that $-\triangle+Q$ is a positive operator, being the sum of two positive operators. The absence of positive eigenvalues was noted in the proof of Lemma 2.10, and it is also known that zero is not an eigenvalue [RS00]. It is thus enough to assume the absence of negative eigenvalues, so-called bound states.

Lemma 2.17. *Assume that $q \in L^\infty_\gamma(\mathbb{R}^3,\mathbb{R})$ with $\gamma > 2$ is such that the Schrödinger operator $H_1 := -\triangle+Q$, $D(H_1) = H^2(\mathbb{R}^3) \subset L^2(\mathbb{R}^3)$, has no negative eigenvalues. Then the energies E_q and E_0 are equivalent,* i.e., *there is a constant $C > 0$ such that*

$$C^{-1}E_0(f_1,f_2) \leq E_q(f_1,f_2) \leq CE_0(f_1,f_2)$$

for all $(f_1,f_2) \in H$.

Proof. See [Phi82, Lemma 3.4] for the proof, which is based on the Gagliardo-Nirenberg-Sobolev inequality (2.13) and the Rellich-Kondrachov compactness theorem. □

We now prove the existence of the wave operator:

Theorem 2.18. *Assume that $q \in L^\infty_\gamma(\mathbb{R}^3,\mathbb{R})$ with $\gamma > 3$ is such that the Schrödinger operator $H_1 := -\triangle+Q$, $D(H_1) = H^2(\mathbb{R}^3) \subset L^2(\mathbb{R}^3)$, has no negative eigenvalues. Let $H_0 = -\triangle$. For*[4] *$j \in \{0,1\}$ set $\mathcal{H}_j = H$ with inner product*[5]

$$\left(\begin{pmatrix} f_1 \\ f_2 \end{pmatrix}, \begin{pmatrix} g_1 \\ g_2 \end{pmatrix}\right)_{\mathcal{H}_j} := (H_j f_1, g_1)_{L^2} + (f_2, g_2)_{L^2}. \tag{2.15}$$

Then the propagators W_j are unitary in $\mathcal{H}_j$, and the wave operator

$$\Omega^\pm = \lim_{t\to\mp\infty} W_1(-t)W_0(t)$$

exists and is an isometry from $\mathcal{H}_0$ to $\mathcal{H}_1$.

Proof. We proceed in the general framework of scattering theory, which is presented, *e.g.*, in [RS79, Sec. XI.10], and which we review here for completeness while proving that the required assumptions are satisfied.

The operators H_j are unbounded positive self-adjoint operators on L^2, with domains $D(H_j) = H^2$; the positivity of H_1 follows from Lemma 2.17 with $f_2 = 0$. We write the free space wave equation ($j = 0$) and plasma wave equation ($j = 1$) as first order systems:

$$\partial_t \begin{pmatrix} u_j(\cdot,t) \\ \partial_t u_j(\cdot,t) \end{pmatrix} = -iA_j \begin{pmatrix} u_j(\cdot,t) \\ \partial_t u_j(\cdot,t) \end{pmatrix}, \qquad A_j = i\begin{pmatrix} 0 & I \\ -H_j & 0 \end{pmatrix}.$$

[4] In the entire section, statements with the index j refer to both cases $j=0$ and $j=1$.

[5] Observe that $((f_1,f_2),(f_1,f_2))_{\mathcal{H}_0} = E_0(f_1,f_2) = \|(f_1,f_2)\|_H^2$ and $((f_1,f_2),(f_1,f_2))_{\mathcal{H}_1} = E_q(f_1,f_2)$. Here, like elsewhere in this text, we identify pairs and column vectors.

The operators A_j are self-adjoint on $D(A_j) = H^2 \oplus H^1 \subset \mathcal{H}_j$:

$$\begin{aligned}\left(A_j\begin{pmatrix} f_1 \\ f_2\end{pmatrix}, \begin{pmatrix} g_1 \\ g_2\end{pmatrix}\right)_{\mathcal{H}_j} &= \left(i\begin{pmatrix} 0 & I \\ -H_j & 0\end{pmatrix}\begin{pmatrix} f_1 \\ f_2\end{pmatrix}, \begin{pmatrix} g_1 \\ g_2\end{pmatrix}\right)_{\mathcal{H}_j} \\ &= (H_j i f_2, g_1)_{L^2} + (-iH_j f_1, g_2)_{L^2} \\ &= (f_2, -iH_j g_1)_{L^2} + (H_j f_1, i g_2)_{L^2} \\ &= \left(\begin{pmatrix} f_1 \\ f_2\end{pmatrix}, A_j\begin{pmatrix} g_1 \\ g_2\end{pmatrix}\right)_{\mathcal{H}_j}\end{aligned}$$

Thus iA_j are skew-adjoint, and consequently the propagators $W_j(t) = e^{-iA_jt}$ are unitary in $\mathcal{H}_j$ [LP67, App. 1, Thm. 2].

Let $B_j = \sqrt{H_j}$, extended to X, which is, again by Lemma 2.17 with $f_2 = 0$, the completion of H^1 in the norm $\|B_j \cdot\|_{L^2} = E_{jq}(\cdot,0)^{1/2} \sim \|\cdot\|_X$.

To analyse the propagators $W_j(t)$, we diagonalize their infinitesimal generators: $A_j = T_j^{-1}\tilde{B}_jT_j$, where

$$T_j = \frac{1}{\sqrt{2}}\begin{pmatrix} B_j & i \\ B_j & -i\end{pmatrix} \qquad T_j^{-1} = \frac{1}{\sqrt{2}}\begin{pmatrix} B_j^{-1} & B_j^{-1} \\ -i & i\end{pmatrix} \qquad \tilde{B}_j = \begin{pmatrix} B_j & 0 \\ 0 & -B_j\end{pmatrix}.$$

The similarity transformations $T_j : \mathcal{H}_j \to L^2 \times L^2$ are isometries, and they also diagonalize the propagators

$$W_j(t) = T_j^{-1}\tilde{W}_j(t)T_j, \qquad \tilde{W}_j(t) = \begin{pmatrix} e^{-iB_jt} & 0 \\ 0 & e^{iB_jt}\end{pmatrix},$$

and the identification operator

$$J := \begin{pmatrix} B_1^{-1}B_0 & 0 \\ 0 & I\end{pmatrix} = T_1^{-1}IT_0 : \mathcal{H}_0 \to \mathcal{H}_1$$

which relates to each other the two Hilbert spaces $\mathcal{H}_0$ and $\mathcal{H}_1$ in which the propagators W_0 and W_1 are unitary.

We shall first prove the existence of the generalized wave operator $\Omega^\pm(A_1,A_0;J) := \lim_{t\to\mp\infty} W_1(-t)JW_0(t)$. As Lemma 2.17 shows, the spaces $\mathcal{H}_0$ and $\mathcal{H}_1$ are actually set-wise, and also topologically, equal, so the identification operator J is in a sense superfluous. Indeed, it turns out that the usual wave operator $\Omega^\pm(A_1,A_0)$ also exists, and that it coincides with $\Omega^\pm(A_1,A_0;J)$; we shall shortly conclude the proof of our theorem by showing this.

Note that the existence of $\Omega^\pm(A_1,A_0;J)$ is equivalent to the existence of $\Omega^\pm(B_1,B_0)$

— for both signs — because as $t \to \mp\infty$,

$$\begin{aligned}
W_1(-t)JW_0(t) &= T_1^{-1}\tilde{W}_1(-t)T_1JT_0^{-1}\tilde{W}_0(t)T_0 \\
&= T_1^{-1}\begin{pmatrix} e^{iB_1t}e^{-iB_0t} & 0 \\ 0 & e^{-iB_1t}e^{iB_0t} \end{pmatrix} T_0 \\
&\xrightarrow{\text{strongly}} T_1^{-1}\begin{pmatrix} \Omega^{\pm}(B_1,B_0) & 0 \\ 0 & \Omega^{\mp}(B_1,B_0) \end{pmatrix} T_0
\end{aligned}$$

if the limits exist. This also shows that if $\Omega^{\pm}(B_1,B_0)$ is an isometry, so is $\Omega^{\pm}(A_1,A_0;J)$.

We show the existence of $\Omega^{\pm}(B_1,B_0)$ using the Kato-Birman invariance principle [RS79, Thm. XI.11]: Let $\lambda > 0$, $T = (0,1/\lambda)$ and $\varphi(s) = (1/s-\lambda)^{1/2}$. Then since $\sigma(H_0) = [0,\infty)$ and $\sigma(H_1) = \sigma_{\mathrm{ac}}(H_1)\cup\sigma_{\mathrm{p}}(H_1)$ with $\sigma_{\mathrm{ac}}(H_1)\subset\sigma_{\mathrm{ess}}(H_1)\subset\sigma_{\mathrm{ess}}(H_0) = [0,\infty)$ [Kat95, Thm. 5.35] and $\sigma_{\mathrm{p}}(H_1)\cap(-\infty,0] = \emptyset$ by assumption, we see that $\sigma((H_j+\lambda)^{-1})\subset[0,1/\lambda] = \overline{T}$. Also, $\varphi' < 0$ on T and $\varphi''\in C(T)\subset L^1_{\mathrm{loc}}(T)$. As for the end points of T, $\varphi(1/\lambda) = 0$ is finite, and $0\notin\sigma_{\mathrm{p}}(H_j+\lambda)$. Therefore, if we succeed in showing that $(H_1+\lambda)^{-1}-(H_0+\lambda)^{-1}$ is trace class, the Kato-Birman invariance principle will imply the existence of $\Omega^{\pm}(B_1,B_0)$, since $B_j = \varphi((H_j+\lambda)^{-1})$.

For showing the trace class property, write $(H_j+\lambda)^{-1} = -R^j_{-\lambda}$. Then

$$\begin{aligned}
(H_1+\lambda)^{-1}-(H_0+\lambda)^{-1} &= R^0_{-\lambda}-R^1_{-\lambda} \\
&= R^0_{-\lambda}\left[-(H_1+\lambda)R^1_{-\lambda}\right]-R^0_{-\lambda}\left[-(H_0+\lambda)\right]R^1_{-\lambda} \\
&= R^0_{-\lambda}\left[H_0+\lambda-H_1-\lambda\right]R^1_{-\lambda} \\
&= -R^0_{-\lambda}QR^1_{-\lambda}
\end{aligned}$$

so that

$$R^1_{-\lambda} = R^0_{-\lambda}+R^0_{-\lambda}QR^1_{-\lambda} = R^0_{-\lambda}\left[I+QR^1_{-\lambda}\right]$$

and thus

$$\begin{aligned}
R^0_{-\lambda}-R^1_{-\lambda} &= -R^0_{-\lambda}QR^0_{-\lambda}\left[I+QR^1_{-\lambda}\right] \\
&= -\left[R^0_{-\lambda}|q|^{1/2}\cdot\right]\left[|q|^{1/2}\operatorname{sgn}q\cdot R^0_{-\lambda}\right]\left[I+QR^1_{-\lambda}\right]
\end{aligned} \tag{2.16}$$

is trace class as a product of two Hilbert-Schmidt operators and a bounded operator: the second factor in (2.16) is of the form $f(x)g(-i\nabla)$, for $f(x) = |q(x)|^{1/2}\operatorname{sgn}q(x)\in L^2$ and $g(\xi) = (|\xi|^2+\lambda)^{-1}\in L^2$, and thus a Hilbert-Schmidt operator [RS79, Thm. XI.20]. For the first factor, note that

$$\left(R^0_{-\lambda}|q|^{1/2}\psi,\eta\right) = \left(|q|^{1/2}\psi,R^0_{-\lambda}\eta\right) = \left(\psi,|q|^{1/2}R^0_{-\lambda}\eta\right),$$

so that $R^0_{-\lambda}|q|^{1/2}\cdot = (|q|^{1/2}\cdot R^0_{-\lambda})^*$. The argument that was used for the second factor therefore shows that the first factor is the adjoint of a Hilbert-Schmidt operator, and thus a Hilbert-Schmidt operator itself, too. The operator (2.16) is thus trace class, which allows us to conclude the existence of the wave operators $\Omega^{\pm}(B_1,B_0)$, and thus also of $\Omega^{\pm}(A_1,A_0;J)$.

To show that the identification operator J can be dropped, apply [RS79, Thm. XI.76] whose assumptions reduce to trivialities except for the following:

(i) Equivalence of the energies $(f,H_j f)$: this again follows from Lemma 2.17 with $f_2 = 0$.

(ii) Vanishing of $\|(H_0-H_1)e^{-iB_0t}w\|_{L^2}$ as $t\to\pm\infty$, for all w in a dense set $D\subset D(H_0) = D(H_1) = H^2(\mathbb{R}^3)$: this will be shown next.

(iii) Existence of the wave operators $\Omega^{\pm}(B_1,V_0)$: this was shown using the Kato-Birman invariance principle.

For proving (ii), choose D to be the set of smooth, quickly decaying functions whose Fourier transform vanishes in a neighbourhood of the origin. Take any $\varepsilon>0$. Then for any $w\in D$,

$$\begin{aligned}\|(H_0-H_1)e^{-iB_0t}w\|_{L^2} &\le \|qe^{-iB_0t}w\|_{L^2(B_R)} + \|qe^{-iB_0t}w\|_{L^2(\mathbb{R}^3\setminus B_R)}\\ &\le \|q\|_{L^\infty}\|e^{-iB_0t}w\|_{L^2(B_R)} + \|q\|_{L^\infty(\mathbb{R}^3\setminus B_R)}\|e^{-iB_0t}w\|_{L^2}.\end{aligned}$$

Now $\|e^{-iB_0t}w\|_{L^2} = (2\pi)^{-3}\|e^{-i|\cdot|t}Fw\|_{L^2} = (2\pi)^{-3}\|Fw\|_{L^2} = \|w\|_{L^2}$, and since q is assumed to decay, we can fix R so large that $\|q\|_{L^\infty(\mathbb{R}^3\setminus B_R)}\|w\|_{L^2} < \varepsilon/2$. It thus suffices to prove that $\|e^{-iB_0t}w\|_{L^2(B_R)}$ goes to zero as $t\to\pm\infty$: For $|x|<R$, integrate by parts in spherical coordinates to see that

$$\begin{aligned}\left|\left(e^{-iB_0t}w\right)(x)\right| &= \frac{1}{(2\pi)^3}\left|\int_{\mathbb{R}^3} e^{ix\cdot\xi}e^{-i|\xi|t}Fw(\xi)\,d\xi\right|\\ &= \frac{1}{(2\pi)^3}\left|\int_{S^2}\int_0^\infty e^{i\rho(x\cdot\theta-t)}Fw(\rho\theta)\rho^2\,d\rho\,dS(\theta)\right|\\ &= \frac{1}{(2\pi)^3}\left|\int_{S^2}\int_0^\infty \frac{ie^{i\rho(x\cdot\theta-t)}}{x\cdot\omega-t}\frac{\partial}{\partial\rho}\left[Fw(\rho\theta)\rho^2\right]\,d\rho\,dS(\theta)\right|\\ &\le \frac{1}{(2\pi)^3}\int_{S^2}\int_0^\infty \frac{|ie^{i\rho(x\cdot\theta-t)}|}{|x\cdot\omega-t|}\left|\frac{\partial}{\partial\rho}Fw(\rho\theta)+\frac{2}{\rho}Fw(\rho\theta)\right|\rho^2\,d\rho\,dS(\theta)\\ &\le \frac{(2\pi)^{-3}}{|t|-R}\int_{\mathbb{R}^3}\left[|\nabla Fw(\xi)|+\frac{2}{|\xi|}|Fw(\xi)|\right]d\xi.\end{aligned}$$

This completes the proof. □

We shall prove shortly in Section 2.4 that the time and frequency domain formulations of the direct scattering problem are equivalent. The following lemma, which shows that the time domain direct scattering problem can be formulated in terms of the wave operator, will allow us to deduce the conservation of energy for the solutions of the direct scattering problem.

Lemma 2.19. *Let* $u_\mathrm{i}(x,t) = \int_{S^2} h(x\cdot\omega - t,\omega)\,dS(\omega)$. *Then* $u_\mathrm{i} \in \tilde{H}$ *if and only if* $h \in L^2(S^2)$, *and* $\tilde{E}_0(u_\mathrm{i},t) = (2\pi)^2\|h\|^2_{L^2(\mathbb{R}\times S^2)}$ *for all* $t \in \mathbb{R}$.

Assume, in addition, that $q \in L^\infty_{\mathrm{comp}}(\mathbb{R}^3,\mathbb{R})$ *is such that the Schrödinger operator* $H_1 := -\Delta + Q$, $D(H_1) = H^2(\mathbb{R}^3) \subset L^2(\mathbb{R}^3)$, *has no negative eigenvalues. Then* $u = u_\mathrm{i} + u_\mathrm{s}$,

$$\begin{pmatrix} u(\cdot,t) \\ \partial_t u(\cdot,t) \end{pmatrix} := W_1(t)\Omega^+ \begin{pmatrix} u_\mathrm{i}(\cdot,0) \\ \partial_t u_\mathrm{i}(\cdot,0) \end{pmatrix},$$

solves (1.7), $u \in \tilde{H}$ *and* $\tilde{E}_q(u,t) = (2\pi)^2\|h\|^2_{L^2(\mathbb{R}\times S^2)}$ *for all* $t \in \mathbb{R}$.

Proof. The isometricity of the incident wave and its translation representation is shown in [LP67, Chapter IV, (2.16)]; there a factor of $1/2\pi$ appears in the expression (1.6) of u_i in terms of h, instead of the definition (1.5) of h. The isometricities of the wave operator Ω^+ and the propagator $W_1(t)$ were shown in Theorem 2.18.

To show that u solves (1.7), first approximate u_i by

$$u_\mathrm{i}^n(x,t) = \int_{S^2} h^n(x\cdot\omega - t,t)\,dS(\omega), \qquad h^n(s,\omega) = 1_{(-\infty,n]}(s)h(s,\omega), \qquad n \in \mathbb{N}.$$

As $n \to \infty$, the isometricities already noted imply that

$$\begin{pmatrix} u_\mathrm{i}^n(\cdot,t) \\ \partial_t u_\mathrm{i}^n(\cdot,t) \end{pmatrix} \to \begin{pmatrix} u_\mathrm{i}(\cdot,t) \\ \partial_t u_\mathrm{i}(\cdot,t) \end{pmatrix}$$

and

$$\begin{pmatrix} u^n(\cdot,t) \\ \partial_t u^n(\cdot,t) \end{pmatrix} := W_1(t)\Omega^+ \begin{pmatrix} u_\mathrm{i}^n(\cdot,0) \\ \partial_t u_\mathrm{i}^n(\cdot,0) \end{pmatrix} \to \begin{pmatrix} u(\cdot,t) \\ \partial_t u(\cdot,t) \end{pmatrix} \tag{2.17}$$

in H for all $t \in \mathbb{R}$. Now u_i^n is $-n$-incoming, *i.e.*, $u_\mathrm{i}^n(x,t) = 0$ when $|x| < -n - t$. Thus

$$\begin{pmatrix} u_\mathrm{i}^n(\cdot,t) \\ \partial_t u_\mathrm{i}^n(\cdot,t) \end{pmatrix} = W_0(t) \begin{pmatrix} u_\mathrm{i}^n(\cdot,0) \\ \partial_t u_\mathrm{i}^n(\cdot,0) \end{pmatrix} = 0$$

in B_R when $t < -R - n$, so that if $\operatorname{supp} q \subset B_R$,

$$W_1(-t)W_0(t) \begin{pmatrix} u_\mathrm{i}^n(\cdot,0) \\ \partial_t u_\mathrm{i}^n(\cdot,0) \end{pmatrix} = W_1(R+n)W_0(-R-n) \begin{pmatrix} u_\mathrm{i}^n(\cdot,0) \\ \partial_t u_\mathrm{i}^n(\cdot,0) \end{pmatrix}$$

when $t < -R - n$. Therefore

$$\begin{pmatrix} u^n(\cdot,t) \\ \partial_t u^n(\cdot,t) \end{pmatrix} = W_1(t+R+n)W_0(-R-n) \begin{pmatrix} u_\mathrm{i}^n(\cdot,0) \\ \partial_t u_\mathrm{i}^n(\cdot,0) \end{pmatrix} = \begin{pmatrix} u_\mathrm{i}^n(\cdot,t) \\ \partial_t u_\mathrm{i}^n(\cdot,t) \end{pmatrix}$$

when $t < -R - n$. Consequently

$$\begin{aligned} \left[\partial_t^2 - \Delta\right] u_\mathrm{s}^n &= -qu^n && \text{in } \mathbb{R}^3 \times \mathbb{R} \\ u_\mathrm{s}^n = \partial_t u_\mathrm{s}^n &= 0 && \text{at } t = -R - n, \end{aligned}$$

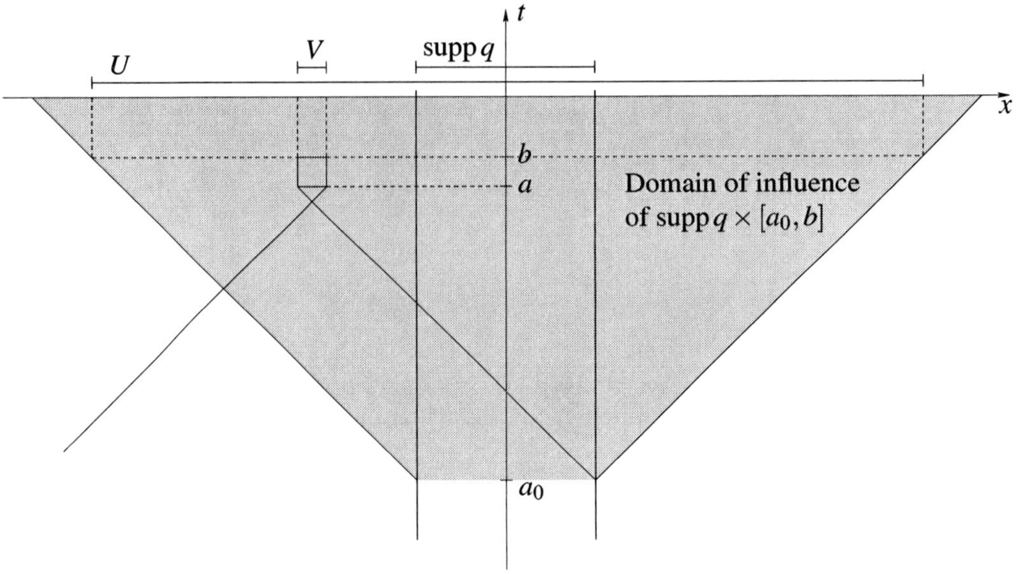

Figure 3: Domains of dependence and influence in (2.18).

and we deduce that $u^n_{\mathrm{s}} = -E_+ * qu^n$.

By (2.17), it suffices to show that as $n \to \infty$, $E_+ * q(u-u^n) \to 0$ uniformly for any compact $V \times [a,b] \subset \mathbb{R}^3 \times \mathbb{R}$. Set $v^n := u - u^n$ and

$$v^n_{\mathrm{s}}(x_0,t_0) := -\int_{-\infty}^{t_0} \frac{1}{4\pi(t_0-t)} \int_{|x-x_0|=t_0-t} q(x)v^n(x,t)\,dS(x)\,dt. \qquad (2.18)$$

By the compactness of the support of q, the integrand vanishes for t smaller than some $a_0 < a$ (see Figure 3). Thus for a large enough compact set $U \subset \mathbb{R}^3$,

$$\begin{aligned} \left[\partial_t^2 - \triangle\right] v^n_{\mathrm{s}} &= -qv^n && \text{in } U \times [a_0,b] \\ v^n_{\mathrm{s}} &= 0 && \text{on } \partial U \times [a_0,b] \\ v^n_{\mathrm{s}} = \partial_t v^n_{\mathrm{s}} &= 0 && \text{at } t = a_0 \end{aligned}$$

and consequently [Eva98, Thm. 7.2.5]

$$\begin{aligned} \operatorname*{ess\,sup}_{a\le t\le b} \left[\|v^n_{\mathrm{s}}(\cdot,t)\|_{H^1_0(V)} + \|\partial_t v^n_{\mathrm{s}}(\cdot,t)\|_{L^2(V)}\right] &\le \operatorname*{ess\,sup}_{a_0\le t\le b} \left[\|v^n_{\mathrm{s}}(\cdot,t)\|_{H^1_0(U)} + \|\partial_t v^n_{\mathrm{s}}(\cdot,t)\|_{L^2(U)}\right] \\ &\le C\|qv^n\|_{L^2(U\times[a_0,b])} \\ &\le C\|q\|_{L^\infty}\left[\int_{a_0}^b \|v^n(\cdot,t)\|_{L^2(U)}\right]^{1/2}. \end{aligned}$$

The quantity on the right hand side approaches zero as $n \to \infty$; to see this, use (2.14), the equivalence of the energies (Lemma 2.17) and the isometricities of the wave operator

(Theorem 2.18) and the translation representation:

$$\|v^n(\cdot,t)\|^2_{L^2(U)} \le C\tilde{E}_0(v^n,t) \le C\tilde{E}_q(v^n,t) = C\tilde{E}_0(v^n_{\mathrm{i}},t) = C\|h\|^2_{L^2(S^2\times(n,\infty))} \xrightarrow{n\to\infty} 0$$

uniformly for all $t \in [a_0,b]$. □

2.4 Equivalence result

We have introduced four different formulations for the direct scattering problem:

(i) $\hat{u} = \hat{u}_{\mathrm{i}} + \hat{u}_{\mathrm{s}}$ satisfies the frequency domain Lippmann-Schwinger equation (1.11)

(ii) $\hat{u} = \hat{u}_{\mathrm{i}} + \hat{u}_{\mathrm{s}}$ satisfies the frequency domain scattering problem (in Section 1.2, page 11)

(iii) $u = u_{\mathrm{i}} + u_{\mathrm{s}}$ satisfies the time domain Lippmann-Schwinger equation (1.7)

(iv) $\big(u(\cdot,t),\partial_t u(\cdot,t)\big) = \Omega^+\big(u(\cdot,t),\partial_t u(\cdot,t)\big)$.

There is a vast literature dealing with each of these four formulations. Each has certain convenient properties that arise naturally in the framework of appropriate function spaces. However, these frameworks are somewhat different for each formulation. In order to be able to combine these properties, we shall now prove the equivalence of the different formulations.

That (iii) follows from (iv) was already shown in Lemma 2.19; if (iii) is uniquely solvable, this also gives the converse implication. The analysis of the others will be broken down into two lemmata showing pairwise equivalences under slightly different assumptions. The results will be summarized in Theorem 2.24.

Lemma 2.20. *Let $k \in \mathbb{R}$ and $\hat{u}_{\mathrm{i}}(\cdot,k) \in L^p_{-\delta}$ for some $\delta > 5/2$ and some $p \in (3,6)$. Assume that $q \in W^{1,\infty}_{\mathrm{loc}}(\mathbb{R}^3,\mathbb{R}) \cap L^\infty_\gamma$ with $\gamma > 2\delta + 3$ is such that the operator $I + \mathcal{G}_0 Q$ is injective on $L^2_{-\delta}$.*

Then $\hat{u}(\cdot,k) = \hat{u}_{\mathrm{i}}(\cdot,k) + \hat{u}_{\mathrm{s}}(\cdot,k)$ satisfies the frequency domain direct scattering problem (in Section 1.2, page 11) if it satisfies the frequency domain Lippmann-Schwinger equation (1.11). For $k \neq 0$, the converse implication also holds.

Proof. Let $\hat{u} = \hat{u}_{\mathrm{i}} + \hat{u}_{\mathrm{s}}$ be a solution of the Lippmann-Schwinger equation (1.11) and let $C_0^\infty(\mathbb{R}^3) \ni \varphi_n \to q\hat{u}$ in L^2_δ (see Lemma 2.3). Then by Lemma 2.4, $\mathcal{G}_k\varphi_n \to \mathcal{G}_k(q\hat{u})$ in $L^p_{-\delta}$, and *a fortiori* in $\mathcal{S}'$. Thus in the sense of distributions,

$$\begin{aligned}(\triangle + k^2)\hat{u} &= (\triangle + k^2)\hat{u}_{\mathrm{s}} \\ &= (\triangle + k^2)\big[-\mathcal{G}_k(q\hat{u})\big] \\ &= \lim_{n\to\infty}\big[-(\triangle + k^2)\Phi\big] * \varphi_n \\ &= \lim_{n\to\infty} \delta_0 * \varphi_n \\ &= q\hat{u}.\end{aligned}$$

To show that the Sommerfeld radiation condition (1.10) is satisfied, we can differentiate under the integral with respect to $r = |x|$, as is seen in the following. Fix $\chi \in C_0^\infty(\mathbb{R}^3)$ with $\operatorname{supp}\chi \subset B_R$. Then

$$\left\langle \frac{\partial}{\partial r}[\Phi * (q\hat{u})], \chi \right\rangle = \left\langle \frac{\partial}{\partial r}[\Phi * (1_{B_R} q\hat{u})], \chi \right\rangle + \left\langle \frac{\partial}{\partial r}\left[\Phi * ((1 - 1_{B_R})q\hat{u})\right], \chi \right\rangle;$$

consider the terms separately. Now as $1_{B_R} q\hat{u} \in \mathcal{E}'$, the convolution in the first term is a usual convolution $\mathcal{E}' \times \mathcal{D}' \to \mathcal{D}'$, and it follows that

$$\begin{aligned}\frac{\partial}{\partial r}\left[\Phi * \left(1_{B_R} q\hat{u}\right)\right](x) &= \sum_{j=1}^{3} \frac{x_j}{|x|} \frac{\partial}{\partial x_j}\left[\Phi * \left(1_{B_R} q\hat{u}\right)\right](x) \\ &= \sum_{j=1}^{3} \frac{x_j}{|x|}\left[\frac{\partial \Phi}{\partial x_j} * \left(1_{B_R} q\hat{u}\right)\right](x) \\ &= \hat{x} \cdot \nabla\Phi * \left(1_{B_R} q\hat{u}\right)(x).\end{aligned}$$

For the second term, we use the mean value theorem: when $x \in \operatorname{supp}\chi \subset B_R$,

$$\begin{aligned}&\frac{\partial}{\partial r}\left[\Phi * \left((1 - 1_{B_R})q\hat{u}\right)\right](x) \\ &= \lim_{h \to 0} \int_{\mathbb{R}^3 \setminus B_R} \left[\frac{\Phi(x + h\hat{x} - y) - \Phi(x - y)}{h}\right] q(y)\hat{u}(y,k)\, dy \\ &= \lim_{h \to 0} \int_{\mathbb{R}^3 \setminus B_R} \hat{x} \cdot \nabla\Phi(x + \xi\hat{x} - y)\, q(y)\hat{u}(y,k)\, dy\end{aligned}$$

for some $\xi = \xi(x,y,h) \in [-h,h]$. The integral converges to the integral of the pointwise limit by Lebesgue's dominated convergence theorem: since $|x| < R$ and $y > R$,

$$|x + \xi\hat{x} - y| \geq |x - y| - |\xi| \geq R - |x| - |h| \geq \frac{R - |x|}{2}$$

when $|h| \leq (R - |x|)/2$, and thus

$$\left|\hat{x} \cdot \nabla\Phi(x + \xi\hat{x} - y)\right| \leq |\hat{x}| \left|\frac{e^{ik|x+\xi\hat{x}-y|}}{4\pi|x + \xi\hat{x} - y|}\right| \left[\left|ik\frac{x + \xi\hat{x} - y}{|x + \xi\hat{x} - y|}\right| + \frac{1}{|x + \xi\hat{x} - y|}\right] \leq C(x,k),$$

showing that the integrand is bounded by the integrable function $C(x,k)|q(y)||\hat{u}(y,k)|$.

Therefore

$$\frac{\partial}{\partial r}[\Phi * (q\hat{u})](x) = \hat{x} \cdot \left[\nabla\Phi * (q\hat{u})\right](x)$$

for $x \in \operatorname{supp}\chi$, and the Sommerfeld radiation condition follows from that for Φ:

$$\begin{aligned}
&\left|\frac{\partial \hat{u}_s}{\partial r}(x,k) - ik\hat{u}(x,k)\right| \\
&\quad = \left|\hat{x}\cdot\left[\nabla\Phi * (q\hat{u})\right](x,k) - ik\left[\Phi * (q\hat{u})\right](x,k)\right| \\
&\quad = \left|\int_{\mathbb{R}^3} \frac{e^{ik|x-y|}}{4\pi|x-y|}\left[\left(\hat{x}\cdot\frac{x-y}{|x-y|} - 1\right)ik - \frac{1}{|x-y|}\right] q(y)\hat{u}(y,k)\,dy\right| \\
&\quad \le Cw(k)\int_{\mathbb{R}^3}\left[\frac{1}{|x-y|}\left|\hat{x}\cdot\frac{x-y}{|x-y|} - 1\right| + \frac{1}{|x-y|^2}\right]|q(y)||\hat{u}(y,k)|\,dy.
\end{aligned}$$

Split the domain of integration into three parts: $B(x,1)$, $B(0,|x|^\alpha)$ for some $\alpha \in (0,1/2)$, and the rest, call it U. In $B(x,1)$, we estimate

$$\left|\hat{x}\cdot\frac{x-y}{|x-y|} - 1\right| \le 2 \tag{2.19}$$

and $|q(y)| \le Cw(y)^{-\gamma} \sim Cw(x)^{-\gamma}$ for large x. In $B(0,|x|^\alpha)$ for $|x|$ large, approximate $\sqrt{1+s} = 1 + s/2 + \mathcal{O}(s^2)$ to get

$$|x-y| = \sqrt{|x|^2 - 2x\cdot y + |y|^2} = |x| - \hat{x}\cdot y + \mathcal{O}\left(\frac{|y|^2}{|x|}\right)$$

and consequently

$$\begin{aligned}
\frac{1}{|x-y|}\left|\hat{x}\cdot\frac{x-y}{|x-y|} - 1\right| + \frac{1}{|x-y|^2} &= \frac{\big||x| - \hat{x}\cdot y - |x-y|\big| + 1}{|x-y|^2} \\
&= \frac{\mathcal{O}\left(\frac{|y|^2}{|x|}\right) + 1}{|x-y|^2} = \frac{\mathcal{O}(1)}{|x-y|^2} = \mathcal{O}\left(\frac{1}{|x|^2}\right).
\end{aligned}$$

In U, simply use (2.19) and $|x-y| \ge 1$. Thus

$$\begin{aligned}
&\left|\frac{\partial \hat{u}_s}{\partial r}(x,k) - ik\hat{u}(x,k)\right| \\
&\quad \le Cw(k)\Bigg[\left\|\frac{1}{|x-y|^2}\right\|_{L^{p'}(B(x,1))} \|w^{\delta-\gamma}\|_{L^\infty(B(x,1))}\|\hat{u}(\cdot,k)\|_{L^p_{-\delta}} \\
&\qquad + \frac{1}{|x|^2}\|w^{\delta-\gamma}\|_{L^{p'}}\|\hat{u}(\cdot,k)\|_{L^p_{-\delta}} + \|w^{\delta-\gamma}\|_{L^{p'}(U)}\|\hat{u}(\cdot,k)\|_{L^p_{-\delta}}\Bigg].
\end{aligned}$$

Now each term is $o(|x|^{-1})$: For the first term,

$$\left\|\frac{1}{|x-y|^2}\right\|_{L^{p'}(B(x,1))} = \left[\int_0^1 r^{2-2p'}\,dr\right]^{1/p'} < \infty$$

since $p' < 3/2$, and $\|\hat{u}(\cdot,k)\|_{L^p} < \infty$ by Corollary 2.14. As $\gamma > \delta + 1$,

$$\|w^{\delta-\gamma}\|_{L^\infty(B(x,1))} \sim w(x)^{\delta-\gamma} = o\left(\frac{1}{|x|}\right).$$

The statement is clear for the second term. For the third one, calculate

$$\begin{aligned}\|w^{\delta-\gamma}\|_{L^{p'}(U)} &= \left[\int_{|x|^\alpha}^{\infty} (1+r)^{p'(\delta-\gamma)} r^2\, dr\right]^{1/p'} \\ &\leq \left[\frac{(1+|x|^\alpha)^{p'(\delta-\gamma)+3}}{p'(\gamma-\delta)-3}\right]^{1/p'} \\ &= \mathcal{O}\left(|x|^{\alpha(\delta-\gamma+\frac{3}{p'})}\right) = o\left(\frac{1}{|x|}\right)\end{aligned}$$

if $\alpha < 1/2$ is chosen large enough. The estimates are uniform over all directions $\hat{x} \in S^2$. Thus $\hat{u}_s$ satisfies the Sommerfeld radiation condition.

To see the converse implication, let $\hat{u} = \hat{u}_i + \hat{u}_s$ be a solution of the frequency domain direct scattering problem at frequency $k \neq 0$. By Lemma 2.15, $\hat{u}_s(\cdot,k) \in H^3_{loc}(\mathbb{R}^3)$, so Green's representation formula can be used to deduce that

$$\begin{aligned}\hat{u}_s(x,k) = &\int_{\partial B_R} \left[\frac{\partial \hat{u}_s}{\partial n}(y)\Phi(x-y) - \hat{u}_s(y)\frac{\partial \Phi}{\partial n(y)}(x-y)\right] dS(y) \\ &- \int_{B_R} \Phi(x-y)q(y)\hat{u}(y,k)\, dy,\end{aligned}$$

which gives the Lippmann-Schwinger equation as $R \to \infty$, since the first integral tends to zero: By assumption,

$$\begin{aligned}&\frac{\partial \hat{u}_s}{\partial n}(y)\Phi(x-y) - \hat{u}_s(y)\frac{\partial \Phi}{\partial n(y)}(x-y) \\ &\qquad = \left[ik\hat{u}_s(y) + o\left(\frac{1}{R}\right)\right]\Phi(x-y) - \hat{u}_s(y)\left[ik\Phi(x-y) + g(x,y)\right]\end{aligned}$$

where $g(x,y) = o(1/R)$ and $\Phi(x-y) = \mathcal{O}(1/R)$ uniformly in all directions, and thus

$$\begin{aligned}&\left|\int_{\partial B_R} \left[\frac{\partial \hat{u}_s}{\partial n}(y)\Phi(x-y) - \hat{u}_s(y)\frac{\partial \Phi}{\partial n(y)}(x-y)\right] dS(y)\right| \\ &\qquad \leq 4\pi R^2 o\left(\frac{1}{R^2}\right) + \left[\int_{\partial B_R} |\hat{u}_s(y)|^2\, dS(y) \int_{\partial B_R} |g(x,y)|^2 dS(y)\right]^{1/2}.\end{aligned}$$

The claim is proved if we show that

$$\int_{\partial B_R} |\hat{u}_s(y)|^2\, dS(y) = \mathcal{O}(1) \qquad \text{as } R \to \infty.$$

To this end, begin by noting that

$$\int_{\partial B_R}\left[\left|\frac{\partial \hat{u}_s}{\partial n}\right|^2+k^2|\hat{u}_s|^2+2k\,\mathrm{Im}\left(\hat{u}_s\frac{\partial \overline{\hat{u}_s}}{\partial n}\right)\right]dS=\int_{\partial B_R}\left|\frac{\partial \hat{u}_s}{\partial n}-ik\hat{u}_s\right|^2 ds \xrightarrow{R\to\infty} 0$$

by the Sommerfeld radiation condition, which was assumed. Thus,

$$\begin{aligned}\lim_{R\to\infty}\int_{\partial B_R}\left[\left|\frac{\partial \hat{u}_s}{\partial n}\right|^2+k^2|\hat{u}_s|^2\right]dS &= -2k\,\mathrm{Im}\lim_{R\to\infty}\int_{\partial B_R}\left(\hat{u}_s\frac{\partial \overline{\hat{u}_s}}{\partial n}\right)dS\\ &= -2k\,\mathrm{Im}\lim_{R\to\infty}\int_{B_R}\left[\hat{u}_s\triangle\overline{\hat{u}_s}+|\nabla\hat{u}_s|^2\right]dx\\ &= -2k\,\mathrm{Im}\lim_{R\to\infty}\int_{B_R}\left[\hat{u}_s(q\overline{\hat{u}}-k^2\overline{\hat{u}_s})+|\nabla\hat{u}_s|^2\right]dx\\ &= -2k\,\mathrm{Im}\int_{\mathbb{R}^3} q\hat{u}_s\overline{\hat{u}}\,dx\end{aligned}$$

and therefore

$$\begin{aligned}\lim_{R\to\infty}\int_{\partial B_R}|\hat{u}_s|^2\,dS &\le \frac{1}{k^2}\lim_{R\to\infty}\int_{\partial B_R}\left[\left|\frac{\partial \hat{u}_s}{\partial n}\right|^2+k^2|\hat{u}_s|^2\right]dS\\ &\le \frac{2}{|k|}\mathrm{Im}\int_{\mathbb{R}^3} Cw^{-\gamma}|\hat{u}_s||\hat{u}|\,dx\\ &\le \frac{2}{|k|}\|\hat{u}_s\|_{L^2_{-\gamma/2}}\|\hat{u}\|_{L^2_{-\gamma/2}}<\infty\end{aligned}$$

by Corollary 2.14. □

Lemma 2.21. *Assume that $q\in L^\infty_\gamma(\mathbb{R}^3,\mathbb{R})$ with $\gamma>2\delta+1$, $\delta>5/2$, is such that the operator $I+\mathcal{G}_0Q$ is injective on $L^2_{-\delta}$. Let $u_i(x,t)=\int_{S^2}h(x\cdot\omega-t,\omega)\,dS(\omega)$, either with $h\in C^2_0(\mathbb{R}\times S^2)$ or with $h\in L^2(\mathbb{R}\times S^2)$ and and q compactly supported. Then the frequency and time domain integral equation formulations (i) and (iii) are equivalent, in the sense of being satisfied for almost all k and t, respectively.*

Proof. First assume that $h\in C^2_0(\mathbb{R}\times S^2)$. Then by using Corollary 2.14 with $m=0$ and both $b=0$ and $b=2$, we see that by assumption, $\hat{u}_i(\cdot,k)$, $\hat{u}_s(\cdot,k)$, $\hat{u}(\cdot,k)\in L^p_{-\delta}$, with norms bounded by $Cw(k)^{-2}$. The inverse Fourier transform of (1.11) gives

$$\begin{aligned}u_s(x,t) &= -\int_{\mathbb{R}}\int_{\mathbb{R}^3}\frac{e^{ik(|x-y|-t)}}{4\pi|x-y|}q(y)\hat{u}(y,k)\,dy\,dk\\ &= -\int_{\mathbb{R}^3}\frac{1}{4\pi|x-y|}q(y)\int_{\mathbb{R}}e^{-ik(t-|x-y|)}\hat{u}(y,k)\,dk\,dy\\ &= -\int_{\mathbb{R}^3}\frac{1}{4\pi|x-y|}q(y)u(y,t-|x-y|)\,dy\\ &= -\left[E_+*(qu)\right](x,t);\end{aligned}$$

changing the order of integration was justified by Fubini's theorem since the integral converges:

$$\begin{aligned}
&\int_{\mathbb{R}}\int_{\mathbb{R}^3}\left|\frac{e^{ik(|x-y|-t)}}{4\pi|x-y|}q(y)\hat{u}(y,k)\right|dy\,dk \\
&\quad\le C\int_{\mathbb{R}}\int_{\mathbb{R}^3}\frac{1}{4\pi|x-y|}w(y)^{-\gamma}|\hat{u}(y,k)|\,dy\,dk \\
&\quad\le C\int_{\mathbb{R}}\left[\int_0^\infty\int_{S^2}r^{-p'}w(x+r\theta)^{p'(\delta-\gamma)}dS(\theta)r^2\,dr\right]^{1/p'}\|\hat{u}(\cdot,k)\|_{L^p_{-\delta}}\,dk<\infty.
\end{aligned}$$

Thus, (i) is equivalent to (iii).

Then consider the case where q is compactly supported. Choose any $\varphi\in C_0^\infty(\mathbb{R}^3)$ and $\psi\in C_0^\infty(\mathbb{R})$. Therefore the integral below converges by Lemmata 2.1 and 2.11, and we can change the order of integration:

$$\begin{aligned}
\langle u_s,\varphi\otimes\psi\rangle &= \langle\hat{u}_s,\varphi\otimes\check{\psi}\rangle \\
&= -\int_{\mathbb{R}^3}\int_{\mathbb{R}}\int_{\mathbb{R}^3}\Phi(x-y,k)q(y)\hat{u}(y,k)\,dy\,\varphi(x)\check{\psi}(k)\,dk\,dx \\
&= -\int_{\mathbb{R}^3}\int_{\mathbb{R}}q(y)\hat{u}(y,k)\int_{\mathbb{R}^3}\Phi(x-y,k)\varphi(x)\,dx\,\check{\psi}(k)\,dk\,dy \\
&= -\lim_{n\to\infty}\int_{\mathbb{R}^3}\int_{\mathbb{R}}q(y)\hat{u}(y,k)\check{\theta}_n(y,k)\,dk\,dy,
\end{aligned}$$

where

$$\check{\theta}_n(y,k)=\chi_n(y)\int_{\mathbb{R}^3}\Phi(x-y,k)\varphi(x)\,dx\,\check{\psi}(k)$$

and $\chi_n\in C_0^\infty(B_n)$ with $\chi_n\equiv 1$ in B_{n-1}. Now

$$\begin{aligned}
\theta_n(y,t) &= \int_{\mathbb{R}}e^{ikt}\chi_n(y)\int_{\mathbb{R}^3}\frac{e^{ik|x-y|}}{4\pi|x-y|}\varphi(x)\,dx\,\check{\psi}(k)\,dk \\
&= \chi_n(y)\int_{\mathbb{R}^3}\frac{1}{4\pi|x-y|}\varphi(x)\int_{\mathbb{R}}e^{ik(|x-y|+t)}\check{\psi}(k)\,dk\,dx \\
&= \chi_n(y)[\mathcal{R}E_+*(\varphi\otimes\psi)](y,t),
\end{aligned}$$

where $\mathcal{R}\cdot$ denotes reflection with respect to time. Therefore as q was assumed compactly supported, we get for a $\chi\in C_0^\infty(\mathbb{R}^3\times\mathbb{R})$ with $\chi\equiv 1$ in a sufficiently large ball, [Hör90, p.104]

$$\begin{aligned}
\langle u_s,\varphi\otimes\psi\rangle &= \lim_{n\to\infty}\langle -q\hat{u},\check{\theta}_n\rangle \\
&= \lim_{n\to\infty}\langle -qu,\theta_n\rangle \\
&= \lim_{n\to\infty}\langle -qu,\chi_n(\chi\mathcal{R}E_+)*(\varphi\otimes\psi)\rangle \\
&= \langle -\chi qu,(\chi\mathcal{R}E_+)*(\varphi\otimes\psi)\rangle \\
&= \langle -(\chi E_+)*(\chi qu),\varphi\otimes\psi\rangle \\
&= \langle -E_+*qu,\varphi\otimes\psi\rangle\,.
\end{aligned}$$

□

The condition of the injectivity of the zero frequency Lippmann-Schwinger operator $I+\mathcal{G}_0Q$ on $L^2_{-\delta}$ has appeared several times both in this section and in Section 2.1. Since formally

$$-\triangle+Q=\triangle(I+\mathcal{G}_0Q),$$

this condition is related to the injectivity of the Schrödinger operator. In the analysis of the time domain formulations in Section 2.3 it was required that the Schrödinger operator has no L^2 eigenvalues, and we noted in particular that zero is not one. In other words, the Schrödinger operator is injective on L^2. However, it might not be injective on a slightly larger space. In this case we talk about resonances:

Definition 2.22. *We say that the Schrödinger operator has a* resonance at zero *if the equation*

$$\varphi=-\mathcal{G}_0Q\varphi$$

has a nontrivial continuous solution uniformly vanishing at infinity.

Now the condition of the injectivity of the Lippmann-Schwinger operator can be formulated in terms of zero resonance:

Lemma 2.23. *Let $\delta>3/2$ and $q\in L^\infty_\gamma$ with $\gamma>\delta+3/2$. Then if the Schrödinger operator has no resonance at zero, the operator $I+\mathcal{G}_0Q$ is injective on $L^2_{-\delta}$.*

Proof. Let $(I+\mathcal{G}_0Q)\varphi=0$ for some $\varphi\in L^2_{-\delta}$. Lemma 2.15 shows that φ is continuous. A simple calculation, similar to the one in the proof of Lemma 3.3 below, shows that φ vanishes uniformly at infinity. Thus $\varphi=0$ by assumption. □

We now summarize the equivalence of the different formulations of the direct scattering problem.

Theorem 2.24. *Let $u_i(x,t)=\int_{S^2}h(x\cdot\omega-t,\omega)\,dS(\omega)$ with $h\in L^2(\mathbb{R}\times S^2)$. Assume that $q\in W^{1,\infty}_{\mathrm{loc}}(\mathbb{R}^3,\mathbb{R})\cap L^\infty_\gamma$ with $\gamma>8$ and that the Schrödinger operator $-\triangle+Q$ has no negative L^2 eigenvalues and no resonance at zero. Then the following are equivalent:*

(i) $\hat{u}(\cdot,k)=\hat{u}_i(\cdot,k)+\hat{u}_s(\cdot,k)$ satisfies the Lippmann-Schwinger equation (1.11) for almost all $k\in\mathbb{R}$

(ii) $\hat{u}(\cdot,k)=\hat{u}_i(\cdot,k)+\hat{u}_s(\cdot,k)$ satisfies the frequency domain scattering problem (in Section 1.2, page 11) for almost all $k\in\mathbb{R}$.

If, in addition, $h\in C^2_0(\mathbb{R}\times S^2)$, (i) and (ii) are equivalent to:

(iii) $u=u_i+u_s$ satisfies the time domain Lippmann-Schwinger equation (1.7).

If q is compatly supported, (i) and (ii) are equivalent to (iii) in the sense of distributions even without the additional assumption about h, and to:

(iv) $\big(u(\cdot,t),\partial_t u(\cdot,t)\big)=W_1(t)\Omega^+\big(u(\cdot,0),\partial_t u(\cdot,0)\big)$ for all $t\in\mathbb{R}$.

Proof. Lemmata 2.1 and 2.11 imply that $\hat{u}(\cdot,k)$, $\hat{u}_{\mathrm{i}}(\cdot,k)$ and $\hat{u}_{\mathrm{s}}(\cdot,k) \in L^2_{-\delta}$ for any $\delta \in (5/2,(\gamma-3)/2)$ and almost all $k \in \mathbb{R}$, so Lemma 2.20 gives the equivalence of the two frequency domain formulations.

The equivalence of the frequency and time domain Lippmann-Schwinger equations was shown in Lemma 2.21. This also guarantees the unique solvability of the time domain Lippmann-Schwinger equation. Therefore the implication of Lemma 2.19 yields the equivalence of the two time domain formulations. □

3 Calculating inner products of solutions from scattering data

Now that we have laid the basis for the analysis of the scattering problem, we are ready to present some new results in this and subsequent sections. In the derivation of these results, we combine properties of the different formulations whose equivalence we have just shown.

As noted in the introduction, a straightforward approach to solving the inverse problem would be to try to simply calculate the potential from the plasma wave equation (1.1):

$$q(x) = \frac{\triangle u(x,t) - \partial_t^2 u(x,t)}{u(x,t)}.$$

The first difficulty with this formula is that we do not know u but only u_{i} and the scattering data. Fortunately, such data allow us to calculate inner products

$$\int_{\mathbb{R}^3} u(x,t)\overline{v(x,s)}\,dx = \lim_{R\to\infty} \int_{B(0,R)} u(x,t)\overline{v(x,s)}\,dx$$

of two solutions u and v through a variant of the Blagoveščenskiĭ identity that we shall now derive. Then if we can localize the solutions into a small neighbourhood of $x_0 \in \mathbb{R}^3$, this inner product with $s=t=t_0$ will give us information about $u(x_0,t_0)$, and varying x_0 and t_0 will permit us to solve the potential.

The Blagoveščenskiĭ identity is a central tool in solving the inverse problem using the boundary control (BC) method. The classical form of this identity allows us to calculate inner products $(u(\cdot,t),v(\cdot,s))$ of two solutions u and v of the plasma wave equation (1.1) in a bounded domain $U \subset \mathbb{R}^n$, if boundary data $\{u|_{\partial U}, \partial_n u|_{\partial U}\}$, $\{v|_{\partial U}, \partial_n v|_{\partial U}\}$ are known [KKL01, Lemma 4.15]. This identity was first derived for the one-dimensional equation [Bla71] and then generalized for the multidimensional case [BB88].

We shall now present a generalization for the case $U = \mathbb{R}^3$, with scattering data instead of boundary data.

3.1 The one-dimensional case

For motivation, we shall first take a look at the one-dimensional case, where the Blagoveščenskiĭ identity for scattering is quite simple: Let u and v be solutions of

$$\left[\partial_t^2 - \partial_x^2 + q(x)\right] u(x,t) = 0 \tag{3.1}$$

$$\left[\partial_t^2 - \partial_x^2 + q(x)\right] v(x,t) = 0. \tag{3.2}$$

Assume that at least one of these solutions, say v, is incoming, *i.e.*, $v(x,t) = 0$ when $|x| \leq a - t$. Define

$$w_R(x,t) = \int_{-R}^{R} u(x,s)\overline{v(x,t)}\,dx, \qquad s,t \in \mathbb{R}, \qquad R > 0.$$

Then using (3.1) and (3.2) and integrating by parts, we see that

$$\begin{aligned}
\left(\partial_s^2 - \partial_t^2\right) w_R(s,t) &= \int_{-R}^{R} \left[\partial_s^2 u(x,s)\overline{v(x,t)} - u(x,s)\overline{\partial_t^2 v(x,t)}\right] dx \\
&= \int_{-R}^{R} \left[\partial_x^2 u(x,s)\overline{v(x,t)} - u(x,s)\overline{\partial_x^2 v(x,t)}\right] dx \\
&= \left[\partial_x u(x,s)\overline{v(x,t)} - u(x,s)\overline{\partial_x v(x,t)}\right]_{x=-R}^{R}.
\end{aligned}$$

We thus have a wave equation in $\mathbb{R}^{1+1}$, with homogeneous initial conditions by the incomingness of v. Therefore

$$\begin{aligned}
w_R(s_0,t_0) = \int_D \Big[&\partial_x u(R,s)\overline{v(R,t)} - u(R,s)\overline{\partial_x v(R,t)} \\
&- \partial_x u(-R,s)\overline{v(-R,t)} + u(-R,s)\overline{\partial_x v(-R,t)}\Big]\, ds\,dt,
\end{aligned} \tag{3.3}$$

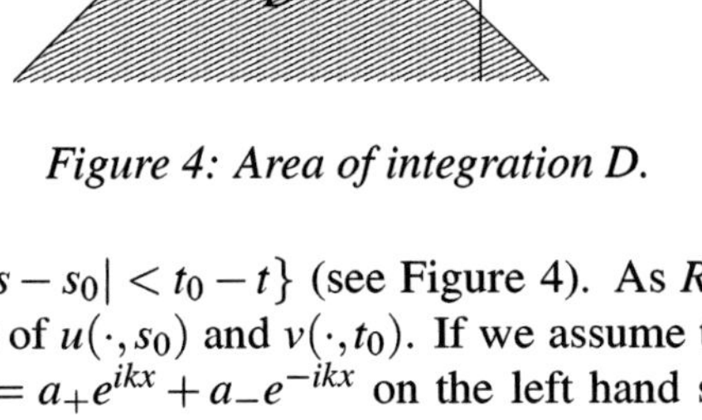

Figure 4: Area of integration D.

where $D = \{(s,t) \in \mathbb{R}^2 \,|\, |s - s_0| < t_0 - t\}$ (see Figure 4). As $R \to \infty$, the left hand side tends to the inner product of $u(\cdot,s_0)$ and $v(\cdot,t_0)$. If we assume the potential q to be compactly supported, $\hat{u}(x,k) = a_+ e^{ikx} + a_- e^{-ikx}$ on the left hand side of the support of the potential and $\hat{u}(x,k) = b_+e^{ikx} + b_-e^{-ikx}$ on the right hand side, and analogously for v. The scattering data are now the relations between the coefficients $a_\pm$ and $b_\pm$, and given $\hat{u}_\mathrm{i}$ and $\hat{v}_\mathrm{i}$, they determine $\hat{u}$ and $\hat{v}$ outside the support of the potential, and through the Fourier transform, also the integrand in (3.3) for R large enough. Plugging this dependence of u and v on the scattering data into (3.3) and taking the limit as $R \to \infty$ yields the one-dimensional Blagoveščenskiĭ identity for scattering.

3.2 The Blagoveščenskiĭ identity

We shall now derive the three-dimensional Blagoveščenskiĭ identity for the scattering problem, which permits us to calculate inner products of two solutions of the plasma wave equation (1.1). We first consider the case where the incident waves u_i and v_i have relatively smooth and compactly supported translation representations. Such incident waves inherit the smoothness, and they are initially incoming and eventually outgoing. For a while, we also assume that we do not only know the scattering amplitude, but also the extended scattering data, *i.e.*, a few of the first terms of the extended far field expansions of scattered waves corresponding to different incident waves.

Theorem 3.1. *Let $q \in L^\infty_\gamma(\mathbb{R}^3,\mathbb{R})$ with $\gamma > 28$ be such that for some $\delta > 5/2$, the operator $I + \mathcal{G}_0 Q$ is injective on $L^2_{-\delta}$. Assume that the incoming incident waves*

$$u_\mathrm{i}(x,t) = \int_{S^2} h_u(x\cdot\omega - t,\omega)\,dS(\omega) \quad \text{and} \quad v_\mathrm{i}(x,t) = \int_{S^2} h_v(x\cdot\omega - t,\omega)\,dS(\omega)$$

are known, with $h_u, h_v \in C_0^\infty(\mathbb{R}\times S^2)$, as well as the corresponding extended scattering data U_j^P, V_j^P, for $P \in \{1, \partial_r, \partial_k^2, \partial_r\partial_k^2\}$ and $j \in \{-1,\ldots,2\}$. Then the inner product

$$\int_{\mathbb{R}^3} u(x,s)\overline{v(x,t)}\,dx$$

can be calculated for all $s,t \in \mathbb{R}$ for which $u(\cdot,s)$ and $v(\cdot,t)$ are square integrable.

Proof. By Corollary 2.14,

$$\begin{aligned}
\|\partial_t^m u_\mathrm{i}(\cdot,t)\|_{L^2_{-\delta}} &= \int_{\mathbb{R}^3} \left| \int_{\mathbb{R}} w(k)^{-a}(-ik)^m e^{-ikt}\hat{u}_\mathrm{i}(x,k)w(k)^a\,dk \right| w(x)^{-2\delta}\,dx \\
&\le \|w^{-2a}\|_{L^2(\mathbb{R})} \int_{\mathbb{R}^3}\int_{\mathbb{R}} |\hat{u}_\mathrm{i}(x,k)|^2 w(k)^{2a+2m}\,dk\,w(x)^{-2\delta}\,dx \\
&\le C\int_{\mathbb{R}}\int_{\mathbb{R}^3} |\hat{u}_\mathrm{i}(x,k)|^2 w(x)^{-2\delta}\,dx\,w(k)^{2a+2m}\,dk \\
&\le C\int_{\mathbb{R}} w(k)^{2(a+m-b)}\,dk < \infty
\end{aligned}$$

if we take $a > 1/2$ and $b > a+m+1/2$. The same estimate applies for u_s. Therefore $\partial_t^m u_\mathrm{i}(\cdot,t), \partial_t^m u_\mathrm{s}(\cdot,t) \in L^2_{-\delta}(\mathbb{R}^3) \subset L^2_\mathrm{loc}(\mathbb{R}^3)$ for all $m \in \mathbb{N}$, uniformly for all $t \in \mathbb{R}$. Define

$$w_R(s,t) := \int_{B_R} u(x,s)\overline{v(x,t)}\,dx.$$

By Green's second identity,

$$
\begin{aligned}
(\partial_s^2-\partial_t^2)w_R(s,t) &= \int_{B_R}\left[\partial_s^2 u(x,s)\overline{v(x,t)} - u(x,s)\partial_t^2\overline{v(x,t)}\right]dx \\
&= \int_{B_R}\left[(\triangle - q(x))u(x,s)\overline{v(x,t)} - u(x,s)\overline{(\triangle - q(x))v(x,s)}\right]dx \\
&= \int_{\partial B_R}\left[\frac{\partial u}{\partial r}(x,s)\overline{v(x,t)} - u(x,s)\overline{\frac{\partial v}{\partial r}(x,t)}\right]dS(x).
\end{aligned}
\tag{3.4}
$$

Fix $s_0, t_0 \in \mathbb{R}$ such that $u(\cdot,s_0)$, $v(\cdot,t_0) \in L^2$. Since v_{i} is an incoming wave, so is v_{s} by the finite speed of propagation, and thus $w_R(s,t) = 0$ when $t \leq -R-a$. Therefore, $w_R(s_0,t_0)$ can be solved from the one-dimensional wave equation (3.4) with homogeneous boundary conditions:

$$
\begin{aligned}
w_R(s_0,t_0) &= 0 + \int_{-\infty}^{t_0}\int_{s_0-t_0+t}^{s_0+t_0-t}\int_{\partial B_R}\left[\frac{\partial u}{\partial r}(x,s)\overline{v(x,t)} - u(x,s)\overline{\frac{\partial v}{\partial r}(x,t)}\right]dS(x)\,ds\,dt \\
&= \int_{\mathbb{R}^2} 1_D(s,t)\int_{\partial B_R}\left[\frac{\partial u}{\partial r}(x,s)\overline{v(x,t)} - u(x,s)\overline{\frac{\partial v}{\partial r}(x,t)}\right]dS(x)\,ds\,dt,
\end{aligned}
$$

where $D = \{(s,t) \in \mathbb{R}^2 \mid |s-s_0| < t_0 - t\}$. Parseval's formula in $\mathbb{R}^2$ now yields the formal integral

$$
w_R(s_0,t_0) = \frac{1}{2\pi}\int_{\mathbb{R}^2}\widehat{1_D}(-\sigma,-\tau)\int_{\partial B_R}\left[\frac{\partial\hat{u}}{\partial r}(x,\sigma)\overline{\hat{v}(x,\tau)} - \hat{u}(x,\sigma)\overline{\frac{\partial\hat{v}}{\partial r}(x,\tau)}\right]dS(x)\,d\sigma\,d\tau.
\tag{3.5}
$$

The part with the solutions $\hat{u}$ and $\hat{v}$ and their partial derivatives behaves nicely by Corollary 2.14, but $\widehat{1_D}$ is not a locally integrable function. We shall now investigate its properties and see how to work with the formal integral (3.5).

Make the change of coordinates $y = s+t$, $z = -s+t$, $\tilde{D} = \{(y,z) \in \mathbb{R}^2 \mid (s,t) \in D\} = (-\infty, t_0+s_0) \times (-\infty, t_0-s_0)$. Consider test functions of the special form $\varphi(\sigma,\tau) = \tilde{\varphi}(\eta,\zeta) = \psi(\eta)\theta(\zeta)$; values of a distribution on test functions of this form define the distribution uniquely [Trè67, Thm. 39.2]. Then

$$
\begin{aligned}
\int_{\mathbb{R}^2}\widehat{1_D}(\sigma,\tau)\varphi(\sigma,\tau)\,d\sigma\,d\tau &= \int_{\mathbb{R}^2} 1_D(s,t)\widehat{\varphi}(s,t)\,ds\,dt \\
&= 2\int_{\mathbb{R}} 1_{(-\infty,t_0+s_0)}(y)\widehat{\psi}(y)\,dy\int_{\mathbb{R}} 1_{(-\infty,t_0-s_0)}(z)\hat{\theta}(z)\,dz.
\end{aligned}
$$

Now as $1_{(-\infty,a)}(y) = \frac{1}{2}[1-\operatorname{sgn}(y-a)]$,

$$
\mathcal{F}\left(1_{(-\infty,a)}\right)(\eta) = \pi\delta_0(\eta) - i\,\mathrm{pv}\frac{e^{ia\eta}}{\eta} =: g(\eta;a)
$$

and thus we get, formally since g is not a locally integrable function,

$$\int_{\mathbb{R}^2} \widehat{1_D}(\sigma,\tau)\varphi(\sigma,\tau)\,d\sigma\,d\tau = 2\int_{-\infty}^{\infty} g(\eta;t_0+s_0)\psi(\eta)\,d\eta \int_{-\infty}^{\infty} g(\zeta;t_0-s_0)\theta(\zeta)\,d\zeta.$$

To turn these formal integrals into actual Lebesgue integrals, integrate by parts:

$$\int_{-\infty}^{\infty} g(\eta;a)\psi(\eta)\,d\eta = -\int_{-\infty}^{\infty} \pi H_0(\eta)\,\psi'(\eta)\,d\eta - i\lim_{\varepsilon\to 0+}\int_{\mathbb{R}\setminus(-\varepsilon,\varepsilon)} \frac{1}{\eta}e^{ia\eta}\psi(\eta)\,d\eta,$$

where H_0 is the Heaviside function. Now

$$\begin{aligned}
&\int_{\mathbb{R}\setminus(-\varepsilon,\varepsilon)} \frac{1}{\eta}e^{ia\eta}\psi(\eta)\,d\eta \\
&\qquad = -\ln|\eta|\,e^{ia\eta}\psi(\eta)\Big|_{\eta=-\varepsilon}^{\varepsilon} - \int_{\mathbb{R}\setminus(-\varepsilon,\varepsilon)} \ln|\eta|\,\left[e^{ia\eta}\psi'(\eta)+iae^{ia\eta}\psi(\eta)\right]\,d\eta \\
&\qquad \xrightarrow{\varepsilon\to 0} -\int_{-\infty}^{\infty} \ln|\eta|\,e^{ia\eta}\psi'(\eta)\,d\eta + \int_{-\infty}^{\infty} ia\int_0^{\eta} \ln|\alpha|\,e^{ia\alpha}\,d\alpha\,\psi'(\eta)\,d\eta.
\end{aligned}$$

Therefore,

$$\int_{-\infty}^{\infty} g(\eta;a)\psi(\eta)\,d\eta = \int_{-\infty}^{\infty} G(\eta;a)\psi'(\eta)\,d\eta$$

where G is the locally integrable function

$$G(\eta;a) = i\ln|\eta|\,e^{ia\eta} + a\int_0^{\eta} \ln|\alpha|\,e^{ia\alpha}\,d\alpha - \pi H_0(\eta),$$

majorised by $\ln|\eta| + a\int_0^{|\eta|}|\ln\alpha|\,d\alpha + \pi$, and consequently

$$\int_{\mathbb{R}^2} \widehat{1_D}(\sigma,\tau)\varphi(\sigma,\tau)\,d\sigma\,d\tau = 2\int_{\mathbb{R}^2} G(\eta;t_0+s_0)G(\zeta;t_0-s_0)\,\partial_\eta\partial_\zeta\tilde{\varphi}(\eta,\zeta)\,d\eta\,d\zeta.$$

Now return to the original variables σ and τ: $\partial_\eta\partial_\zeta = \frac{1}{4}(\partial_\tau^2 - \partial_\sigma^2)$ and thus

$$\int_{\mathbb{R}^2} \widehat{1_D}(\sigma,\tau)\varphi(\sigma,\tau)\,ds\,dt = \int_{\mathbb{R}^2} F(t_0,s_0,\tau,\sigma)\left[\frac{\partial^2\varphi}{\partial\tau^2} - \frac{\partial^2\varphi}{\partial\sigma^2}\right]\,d\sigma\,d\tau,$$

where F is the locally integrable function

$$F(s_0,t_0,\sigma,\tau) = \frac{1}{4}G(\tau+\sigma,t_0+s_0)G(\tau-\sigma;t_0-s_0). \tag{3.6}$$

In order to express

$$\begin{aligned}
w_R(s_0,t_0) = \frac{1}{2\pi}\int_{\mathbb{R}^2} F(-\sigma,-\tau)\int_{\partial B_R} \Bigg[&\frac{\partial\hat{u}}{\partial r}(x,\sigma)\overline{\frac{\partial^2\hat{v}}{\partial\tau^2}(x,\tau)} - \frac{\partial^3\hat{u}}{\partial\sigma^2\partial r}(x,\sigma)\overline{\hat{v}(x,\tau)} \\
&- \hat{u}(x,\sigma)\overline{\frac{\partial^3\hat{v}}{\partial\tau^2\partial r}(x,\tau)} + \frac{\partial^2\hat{u}}{\partial\sigma^2}(x,\sigma)\overline{\frac{\partial\hat{v}}{\partial r}(x,\tau)}\Bigg]\,dS(x)\,d\sigma\,d\tau
\end{aligned} \tag{3.7}$$

in terms of the known incoming waves $\hat{u}_{\mathrm{i}}$ and $\hat{v}_{\mathrm{i}}$ and the far field data for the corresponding scattering solutions $\hat{u}_{\mathrm{s}}$ and $\hat{v}_{\mathrm{s}}$, we use the expansions given by Lemma 3.3 below:

$$
\begin{aligned}
&\int_{\mathbb{R}^2} F(-\sigma,-\tau)\int_{\partial B_R}\frac{\partial\hat{u}}{\partial r}(x,\sigma)\overline{\frac{\partial^2\hat{v}}{\partial\tau^2}(x,\tau)}\,dS(x)\,d\sigma\,d\tau\\
&\quad=\int_{\mathbb{R}^2} F(-\sigma,-\tau)\int_{\partial B_R}\left[\frac{\partial\hat{u}_{\mathrm{i}}}{\partial r}(x,\sigma)+e^{i\sigma R}\sum_{j=1}^{N}\frac{U_j^{\partial_r}(\hat{x},\sigma)}{R^j}+E_N^{\partial_r\hat{u}}(x,\sigma)\right]\times\\
&\qquad\times\overline{\left[\frac{\partial^2\hat{v}_{\mathrm{i}}(x,\tau)}{\partial\tau^2}+e^{i\tau R}\sum_{j=-1}^{N}\frac{V_j^{\partial_k^2}(\hat{x},\tau)}{R^j}+E_N^{\partial_k^2\hat{v}}(x,\tau)\right]}\,dS(x)\,d\sigma\,d\tau,
\end{aligned}
\tag{3.8}
$$

and similarly for the three other terms in (3.7).

We now postpone the proof Theorem 3.1 until we have developed these expansions. After this, in Section 3.3 we shall see that our assumptions guarantee that the terms involving the remainders tend to zero as $R=|x|\to\infty$. On the other hand, since the inner product $w(x_0,t_0):=\lim_{R\to\infty}w_R(s_0,t_0)$ is assumed finite, the terms that grow as $R\to\infty$ must cancel each other. Thus, we shall be able to calculate $w(s_0,t_0)$ in terms of $\hat{u}_{\mathrm{i}}$, $\hat{v}_{\mathrm{i}}$, U_j^P and V_j^P.

3.3 Extended far field expansion

For completing the proof of Theorem 3.1, we derive the expansions used in (3.8) and its analogues, and a series of estimates for the remainders. The expansion will follow by plugging the following expansions of the fundamental solution into the Lippman-Schwinger equation (1.11):

Lemma 3.2. *Let $N \in \mathbb{N}$ and $0 < \alpha < 1$. The fundamental solution $\Phi(x-y,k) = \frac{e^{ik|x-y|}}{4\pi|x-y|}$ to the Helmholtz equation and its partial derivatives have the asymptotics*

$$\Phi(x-y,k) = \sum_{j=1}^{N} \frac{e^{ik|x|}}{|x|^j} e^{-ik\hat{x}\cdot y} \sum_{m=0}^{N-1} \sum_{l=0}^{2j-2} \sum_{s=0}^{2j-2-l} a_{jlms} k^m (\hat{x}\cdot y)^l |y|^s + g_N(x,y,k) \tag{3.9}$$

$$\frac{\partial \Phi}{\partial k}(x-y,k) = \sum_{j=0}^{N-1} \frac{e^{ik|x|}}{|x|^j} e^{-ik\hat{x}\cdot y} \sum_{m=0}^{N-1} \sum_{l=0}^{2j} \sum_{s=0}^{2j-l} a^k_{jlms} k^m (\hat{x}\cdot y)^l |y|^s + g^k_N(x,y,k)$$

$$\frac{\partial^2 \Phi}{\partial k^2}(x-y,k) = \sum_{j=-1}^{N-1} \frac{e^{ik|x|}}{|x|^j} e^{-ik\hat{x}\cdot y} \sum_{m=0}^{N} \sum_{l=0}^{2j+2} \sum_{s=0}^{2j+2-l} a^{kk}_{jlms} k^m (\hat{x}\cdot y)^l |y|^s + g^{kk}_N(x,y,k)$$

$$\frac{\partial \Phi}{\partial r}(x-y,k) = \sum_{j=1}^{N+2} \frac{e^{ik|x|}}{|x|^j} e^{-ik\hat{x}\cdot y} \sum_{m=0}^{N} \sum_{l=0}^{2j-2} \sum_{s=0}^{2j-2-l} a^r_{jlms} k^m (\hat{x}\cdot y)^l |y|^s + g^r_N(x,y,k)$$

$$\frac{\partial^2 \Phi}{\partial r \partial k}(x-y,k) = \sum_{j=0}^{N} \frac{e^{ik|x|}}{|x|^j} e^{-ik\hat{x}\cdot y} \sum_{m=1}^{N} \sum_{l=0}^{2j+2} \sum_{s=0}^{2j+2-l} a^{rk}_{jlms} k^m (\hat{x}\cdot y)^l |y|^s + g^{rk}_N(x,y,k)$$

$$\frac{\partial^3 \Phi}{\partial r \partial k^2}(x-y,k) = \sum_{j=-1}^{N-1} \frac{e^{ik|x|}}{|x|^j} e^{-ik\hat{x}\cdot y} \sum_{m=0}^{N} \sum_{l=0}^{2j+2} \sum_{s=0}^{2j+2-l} a^{rkk}_{jlms} k^m (\hat{x}\cdot y)^l |y|^s + g^{rkk}_N(x,y,k)$$

for some constants a_{jlms}, a^k_{jlms}, a^{kk}_{jlms}, a^r_{jlms}, a^{rk}_{jlms}, a^{rkk}_{jlms}. The remainders satisfy

$$\begin{aligned}
|g_N(x,y,k)| &\le Cw(k)^N w(x)^{N(2\alpha-1)-1} \\
|g^k_N(x,y,k)| &\le Cw(k)^N w(x)^{N(2\alpha-1)} \\
|g^{kk}_N(x,y,k)| &\le Cw(k)^N w(x)^{N(2\alpha-1)+1} \\
|g^r_N(x,y,k)| &\le Cw(k)^{N+1} w(x)^{(N+3)(2\alpha-1)-2\alpha} \\
|g^{rk}_N(x,y,k)| &\le Cw(k)^{N+1} w(x)^{(N+1)(2\alpha-1)+2\alpha} \\
|g^{rkk}_N(x,y,k)| &\le Cw(k)^{N+1} w(x)^{(N+1)(2\alpha-1)+2\alpha}
\end{aligned}$$

uniformly for all $|y| \le |x|^\alpha \to \infty$, $k \in \mathbb{R}$.

Proof. Let $|y| \le |x|^\alpha$ and consider the behaviour as $|x| \to \infty$. Use the Maclaurin expansions for $(1+s)^{\pm 1/2}$ to see that

$$\begin{aligned}
|x-y|^{\pm 1} &= |x|^{\pm 1} \left(1 - 2\frac{\hat{x}\cdot y}{|x|} + \frac{|y|^2}{|x|^2} \right)^{\pm 1/2} \\
&= |x|^{\pm 1} \left(1 + \sum_{j=1}^{\infty} c^{\pm}_j \sum_{l=0}^{j} \binom{j}{l} \frac{[2(\hat{x}\cdot y)]^{j-l} |y|^{2l}}{|x|^{j+l}} \right) \qquad (3.10) \\
&= |x|^{\pm 1} + \sum_{j=1}^{N} \sum_{m=0}^{\lfloor j/2 \rfloor} c^{\pm}_{jm} \frac{(\hat{x}\cdot y)^{j-2m} |y|^{2m}}{|x|^{j\mp 1}} + \mathcal{O}\left(|x|^{(N+1)(\alpha-1)\pm 1} \right),
\end{aligned}$$

with $c_{j0}^{+} = 1$. Plug this with the + sign and $N = 1$ into the Maclaurin expansion of e^{it}: denoting the remainder in (3.10) by $g(x,y) = \mathcal{O}(|x|^{2\alpha-1})$, this gives

$$
\begin{aligned}
e^{ik|x-y|} &= e^{ik|x|}e^{-ik\hat{x}\cdot y}\left[\sum_{m=0}^{N-1}\frac{(ikg(x,y))^m}{m!} + \mathcal{O}\left(|kg(x,y)|^N\right)\right] \\
&= e^{ik|x|}e^{-ik\hat{x}\cdot y}\left[\sum_{m=0}^{N-1}\sum_{j=0}^{N-1}\sum_{l=0}^{2j}\sum_{s=0}^{2j-l} c_{jmls}k^m\frac{(\hat{x}\cdot y)^l|y|^p}{|x|^j} + \mathcal{O}\left(|x|^{N(2\alpha-1)}\right)\right].
\end{aligned}
$$

Multiplying these expansions gives the statement. □

Lemma 3.3. *Let $\delta \geq 0$, $N \in \mathbb{N}$, $3 < p \leq \infty$ and $0 < \alpha < 1$. Assume that $q \in L_\gamma^\infty$ and that γ and N are so large that the exponents M_N^P, $P \in \{1, \partial_r, \partial_k^2, \partial_r\partial_k^2\}$, in (3.19)–(3.22) below are negative. Let $\hat{u} = \hat{u}_{\mathrm{i}} + \hat{u}_{\mathrm{s}}$ be such that the norms in (3.15)–(3.18) are finite. Then the scattering solution $\hat{u}_{\mathrm{s}}$ of the Lippman-Schwinger equation (1.11) and its partial derivatives admit the extended far field expansions*

$$\hat{u}_{\mathrm{s}}(x,k) = e^{ik|x|}\sum_{j=1}^{N}\frac{1}{|x|^j}U_j^1(\hat{x},k) + E_N^1(x,k) \tag{3.11}$$

$$\frac{\partial^2\hat{u}_{\mathrm{s}}}{\partial k^2}(x,k) = e^{ik|x|}\sum_{j=-1}^{N}\frac{1}{|x|^j}U_j^{\partial_k^2}(\hat{x},k) + E_N^{\partial_k^2}(x,k) \tag{3.12}$$

$$\frac{\partial\hat{u}_{\mathrm{s}}}{\partial r}(x,k) = e^{ik|x|}\sum_{j=1}^{N}\frac{1}{|x|^j}U_j^{\partial_r}(\hat{x},k) + E_N^{\partial_r}(x,k) \tag{3.13}$$

$$\frac{\partial^3\hat{u}_{\mathrm{s}}}{\partial r\partial k^2}(x,k) = e^{ik|x|}\sum_{j=-1}^{N}\frac{1}{|x|^j}U_j^{\partial_r\partial_k^2}(\hat{x},k) + E_N^{\partial_r\partial_k^2}(x,k) \tag{3.14}$$

where

$$|E_N^1(x,k)| \leq Cw(k)^N w(x)^{M_N^1}\|\hat{u}(\cdot,k)\|_{L^2_{-\delta}} \tag{3.15}$$

$$|E_N^{\partial_k^2}(x,k)| \leq Cw(k)^N w(x)^{M_N^{\partial_k^2}}\left[\|\hat{u}(\cdot,k)\|_{L^2_{-\delta}} + \|\partial_k\hat{u}(\cdot,k)\|_{L^2_{-\delta}} + \|\partial_k^2\hat{u}(\cdot,k)\|_{L^2_{-\delta}}\right] \tag{3.16}$$

$$|E_N^{\partial_r}(x,k)| \leq Cw(k)^{N+1} w(x)^{M_N^{\partial_r}}\|\hat{u}(\cdot,k)\|_{L^p_{-\delta}} \tag{3.17}$$

$$|E_N^{\partial_r\partial_k^2}(x,k)| \leq Cw(k)^{N+1} w(x)^{M_N^{\partial_r\partial_k^2}}\left[\|\hat{u}(\cdot,k)\|_{L^2_{-\delta}} + \|\partial_k\hat{u}(\cdot,k)\|_{L^2_{-\delta}} + \|\partial_k^2\hat{u}(\cdot,k)\|_{L^p_{-\delta}}\right] \tag{3.18}$$

as $x \to \infty$, uniformly in all directions, with

$$M_N^1 = \max\left\{\alpha(\delta-\gamma+\tfrac{3}{2}), N(2\alpha-1)-1, \alpha(\delta-\gamma+2N-\tfrac{3}{2})-1\right\} \tag{3.19}$$
$$M_N^{\partial_k^2} = \max\left\{N(2\alpha-1)+1, \alpha(\delta-\gamma+2N-1/2)+1\right\} \tag{3.20}$$
$$M_N^{\partial_r} = \max\left\{\alpha(\delta-\gamma+2N+1-\tfrac{3}{p}), \alpha(\delta-\gamma)+3-\tfrac{3}{p}, (N+3)(2\alpha-1)-2\alpha\right\} \tag{3.21}$$
$$M_N^{\partial_r\partial_k^2} = \max\left\{\alpha(\delta-\gamma+2N+1-\tfrac{3}{p}), \alpha(\delta-\gamma)+3-\tfrac{3}{p}, (N+1)(2\alpha-1)+2\alpha\right\}. \tag{3.22}$$

Remark. A few of the last terms in (3.11)–(3.14) actually decay faster in x than the error terms E_N^P. This notation, however, simplifies the indices.

Proof. For $\hat{u}_s$, work directly on the expression

$$\hat{u}_s(x,k) = -\int_{\mathbb{R}^3} \Phi(x-y,k)q(y)\hat{u}(y,k)\,dy.$$

The calculations in the proof of Lemma 2.20 justify differentiation under the integral for $\partial_r\hat{u}_s$. For $\partial_k^2\hat{u}_s$, $\partial_r\hat{u}_s$ and $\partial_r\partial_k^2\hat{u}_s$, the argument is similar, only simpler since the integrand is less singular. When differentiating with respect to k, the Leibniz rule gives three terms.

Split the domain of integration into three parts: $B(0,|x|^\alpha)$, $(\mathbb{R}^3\setminus B(0,|x|^\alpha))\cap B(x,1) =: U_1$ and $\mathbb{R}^3\setminus(B(0,|x|^\alpha)\cup B(x,1)) =: U_2$. (See Figure 5.)

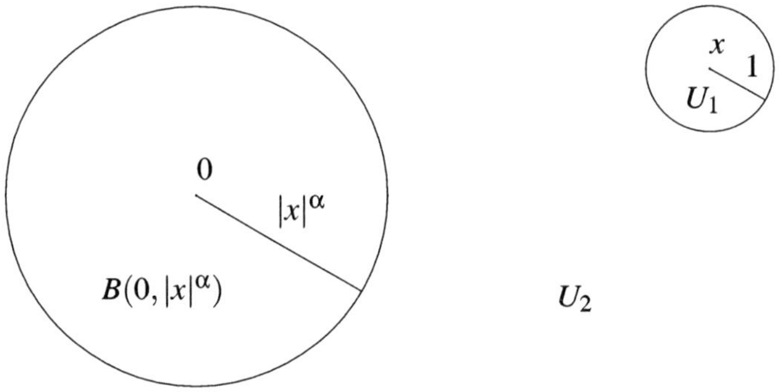

Figure 5: Three domains of integration $B(0,|x|^\alpha)$, U_1 and U_2.

In the first part, use the expansions of Lemma 3.2. Its main terms give for $\hat{u}_s$, as $|x|\to\infty$, the main part of the expansions (3.11)–(3.14) plus some error terms with x dependence of the type $w(x)^{\alpha(\delta-\gamma+2N-1/2)-1}$. The integrals of the remainder terms are estimated using the Hölder inequality with weights,

$$\left|\int f(x)g(x)\,dx\right| \le \|f\|_{L^r_\delta}\|g\|_{L^{r'}_{-\delta}}, \qquad \frac{1}{r}+\frac{1}{r'}=1, \tag{3.23}$$

now with $r=2$, yielding terms with x dependence of the type $w(x)^{N(2\alpha-1)-1}$.

In U_1, $\Phi(x-y)$ has a square integrable singularity at $y=x$, uniform as $|x| \to \infty$. Thus the decay of q yields an estimate for the integral over U_1, with x dependence of the type $w(x)^{2\alpha(\delta-\gamma+3/2)}$. The same is true for $\partial_k\Phi$, $\partial_k^2\Phi$, $\partial_r\partial_k\Phi$ and $\partial_r\partial_k^2\Phi$. However,

$$\partial_r\Phi(x-y,k) = \frac{e^{ik|x-y|}}{4\pi|x-y|}\left(ik - \frac{1}{|x-y|}\right)\hat{x}\cdot\frac{x-y}{|x-y|}$$

has a stronger singularity which is integrable only to powers less than $3/2$. For estimating its integral, we can use (3.23) with $r > 3$ (*i.e.*, $r' < 3/2$).

Finally in U_2, Φ is bounded, so the decay of q gives an estimate of the integral over U_2, with x dependence of the type $w(x)^{\alpha(\delta-\gamma+3/2)}$. The integrals of the partial derivatives of Φ can be estimated in a similar way; $\partial_k^2\Phi(x-y,k)$ and $\partial_r\partial_k^2\Phi(x-y,k)$ grow as $|x-y| \sim |x|$, but this only gives a slightly slower decay for the estimate.

Combining these estimates completes the proof. □

We now have enough tools to finish the proof of the Blagoveščenskiĭ identity:

Proof of Theorem 3.1, continued. Estimate the error terms in (3.8) and its analogues using Lemma 3.3 and Corollary 2.14. The smoothness assumption about h_u and h_v guarantees the convergence of the integral with respect to σ and τ. The assumption about γ shows that if we choose[6] $\alpha = 1/4$ and $N = 7$ in Lemma 3.3, the integrals involving the error terms tend to zero as $R \to \infty$. All other terms are assumed known. □

3.4 Sufficiency of regular scattering data

When deriving the Blagoveščenskiĭ identity in the previous sections, we assumed that we know the *extended* far field data, *i.e.*, the first few terms of the far field expansions of $\hat{u}_s$, $\partial_r\hat{u}_s$, $\partial_k^2\hat{u}_s$ and $\partial_r\partial_k^2\hat{u}_s$. It turns out that these extended data are actually redundant: they are determined by the regular far field data $\hat{u}_s^\infty = U_1^1$. For the case of a compactly supported potential, this is very easy to see:

Lemma 3.4. *Assume that q is compactly supported. Then the far field pattern $\hat{u}_s^\infty = U_1^1$ determines the extended far field data U_j^P for $j \in \{-1,0,1,\ldots\}$ and $P \in \{1, \partial_k^2, \partial_r, \partial_r\partial_k^2\}$.*

Proof. Since $\hat{u}_s$ satisfies the Helmholtz equation outside the support of q, Lemma 2.15 shows that $\hat{u}$ is as many times continuously differentiable as we want sufficiently far from $\operatorname{supp} q$. By Rellich's lemma [CK98, Thm. 2.13], the far field data determine $\hat{u}_s(x,k)$ uniquely for large $|x|$. This clearly also fixes the lower order asymptotics U_j^P. □

Remark. The scattered wave $\hat{u}_s$ can actually be determined constructively in the following way:[7] Consider the spherical harmonic expression

$$\hat{u}_s(r\hat{x},k) = \sum_{n=0}^{\infty}\sum_{m=-n}^{n} a(r,k)Y_n^m(\hat{x}).$$

[6]This choice of α is optimal for $\hat{u}_s$. For the others, this uniform choice simplifies the calculations.

[7]This construction is, however, very unstable, since for fixed r, the spherical Bessel functions grow rapidly, $h_n^{(1)}(kr) \sim n^n$.

The coefficients a_n^m are given by

$$a_n^m(r,k) = \int_{S^2} \hat{u}_s(r\hat{x},k)\overline{Y_n^m(\hat{x})}\,dS(x) \tag{3.24}$$

Differentiating under the integral, integrating by parts and using properties of the spherical harmonics we see that the coefficients satisfy the spherical Bessel differential equation

$$\left[\frac{d^2}{dr^2} + \frac{2}{r}\frac{d}{dr} + \left(k^2 - \frac{n(n+1)}{r^2}\right)\right] a_n^m(r,k) = 0$$

and thus $a_n^m(k,r) = c_n^m(k)h_n^{(1)}(kr) + d_n^m(k)h_n^{(2)}(kr)$. The Sommerfeld radiation condition forces $d_n^m \equiv 0$, and thus

$$a_n^m(k,r) = c_n^m(k)h_n^{(1)}(kr) = c_n^m(k)\frac{(-1)^n}{i}\frac{e^{ikr}}{kr} + \mathcal{O}\left(\frac{1}{r^2}\right)$$

On the other hand, inserting the far field expansion, which is uniform in all directions $\hat{x}$, into (3.24) gives

$$a_n^m(r,k) = \frac{e^{ikr}}{r}\int_{S^2} \hat{u}_s^\infty(\hat{x},k)\overline{Y_n^m(\hat{x})}\,dS(x) + \mathcal{O}\left(\frac{1}{r^2}\right).$$

Equating these two expressions shows that the coefficients c_n^m are completely determined by the far field data. Thus so too are a_n^m, and hence the whole scattered wave $\hat{u}_s(x,k)$ for large $|x|$.

The regular far field pattern $\hat{u}_s^\infty = U_1^1$ also determines the lower order terms U_j^P when the potential is not compactly supported but decays sufficiently fast. This will be stated soon in Theorem 3.9. We shall use different techniques to show this for $P = 1$, $P = \partial_r$, and $P \in \{\partial_k^2, \partial_r\partial_k^2\}$. For this reason, the proof of Theorem 3.9 will be broken down into a series of lemmata.

Lemma 3.5. *Let $\delta > 5/2$, $3 < p \leq 6$, $N \in \mathbb{N}$ and $0 < \alpha < 1$, with $q \in W_\gamma^{1,\infty}(\mathbb{R}^3,\mathbb{R})$ such that the operator $I + \mathcal{G}_0 Q$ is injective on $L^2_{-\delta}$. Assume that γ and N are so large that the exponents in (3.25) are negative. Let $u_i(x,t) = \int_{S^2} h(x\cdot\omega - t,\omega)\,dS(\omega)$ be an incident free space wave with $h \in C_0^\infty(\mathbb{R}\times S^2)$. Then the second radial partial derivative of the scattering solution $\hat{u}_s$ of the Lippman-Schwinger equation (1.11) admits the extended far field expansion*

$$\frac{\partial^2 \hat{u}_s}{\partial r^2}(x,k) = e^{ik|x|}\sum_{j=1}^{N}\frac{1}{|x|^j}U_j^{\partial_r^2}(\hat{x},k) + E_N^{\partial_r^2}(x,k),$$

where

$$|E_N^{\partial_r^2}(x,k)| \leq Cw(k)^{N+1}w(x)^{M_N^{\partial_r^2}}\left[\|\hat{u}(\cdot,k)\|_{L^p_{-\delta}} + \|\nabla\hat{u}(\cdot,k)\|_{L^p_{-\delta}}\right] \tag{3.25}$$

as $x\to\infty$, uniformly in all directions, with $M_N^{\partial_r^2} = M_N^{\partial_r}$ defined in (3.21) in Lemma 3.3.

Proof. Differentiate under the integral, as in the proof of Lemma 3.3, and integrate by parts to get

$$\begin{aligned}\partial_r^2 \hat{u}_s(x,k) &= -\partial_r \int_{\mathbb{R}^3} \hat{x}\cdot\nabla_x\Phi(x-y,k)q(y)\hat{u}(y,k)\,dy \\ &= -\partial_r \int_{\mathbb{R}^3} \Phi(x-y,k)\hat{x}\cdot\nabla\left[q(y)\hat{u}(y,k)\right]dy \\ &= -\int_{\mathbb{R}^3} \hat{x}\cdot\nabla_x\Phi(x-y,k)\hat{x}\cdot\nabla\left[q(y)\hat{u}(y,k)\right]dy.\end{aligned} \tag{3.26}$$

The integral converges since

$$\begin{aligned}\|\nabla\left[q(y)\hat{u}(y,k)\right]\|_{L^p_{-\delta}} \le \|\nabla q\|_{L^\infty}\|\hat{u}_i(\cdot,k)\|_{L^p_{-\delta}} + \|\nabla q\|_{L^\infty}\|\hat{u}_s(\cdot,k)\|_{L^p_{-\delta}} \\ + \|q\|_{L^\infty}\|\nabla\hat{u}_i(\cdot,k)\|_{L^p_{-\delta}} + \|q\|_{L^\infty}\|\nabla\hat{u}_s(\cdot,k)\|_{L^p_{-\delta}}\end{aligned}$$

and the norms on the right hand side are finite: Clearly $\|q\|_{L^\infty}$, $\|\nabla q\|_{L^\infty} \le \|q\|_{W^{1,\infty}_\gamma} < \infty$ directly by the assumptions. The finiteness of the norms of $\hat{u}_i$ and $\hat{u}_s$ was shown in Corollary 2.14. Since

$$\partial_{x_j}\hat{u}_i(x,k) = \mathcal{F}\left[\int_{S^2} \omega_j h'(x\cdot\omega-\cdot,\omega)\,dS(\omega)\right],$$

the proof of Lemma 2.2 guarantees the finiteness of the norm of $\nabla\hat{u}_i$. For the norm of $\nabla\hat{u}_s$, argue as follows: Simply leaving out the dot product with $\hat{x}$ in the proof of Lemma 3.3 and using the fact that $\hat{u}_s \in L^p_{-\delta}$ for $\delta > 2$, which we already know, we see that $\nabla\hat{u}_s(x,k) = \mathcal{O}(|x|^{-1})$. By Lemma 2.15, $\hat{u}_s(\cdot,k) \in C^1$ and thus $\nabla\hat{u}_s(\cdot,k)$ is also locally bounded. Therefore $\nabla\hat{u}_s(\cdot,k) \in L^2_{-\delta}$.

We can now plug the expansion in Lemma 3.2 for $\partial_r\Phi$ into (3.26). The claim of the theorem is proved in the same way as the one about $\partial_r\hat{u}_s$ in Lemma 3.3, but now with $\hat{x}\cdot\nabla[q(y)\hat{u}(y,k)]$ under the integral instead of just $q(y)\hat{u}(y,k)$. □

Remark. We get the same speed of decay in x since we assumed more smoothness of q and integrated by parts. The same could have been done in Lemma 3.3 to get $M_N^{\partial_r} = M_N^1$ and $M_N^{\partial_r\partial_k^2} = M_N^{\partial_k^2}$. This would also have got rid of the $L^p_{-\delta}$ norms, $p > 3$. Assuming $q \in W^{2,\infty}_\gamma$, we could have integrated by parts twice in the proof of Lemma 3.5 to get $M_N^{\partial_r^2} = M_N^1$.

Lemma 3.6. *Let $q \in W^{1,\infty}_\gamma(\mathbb{R}^3,\mathbb{R})$, $\gamma > 24\frac{1}{2}$ be such that the operator $I + \mathcal{G}_0 Q$ is injective on $L^2_{-\delta}$ for some $\delta > 5/2$. Assume that $\hat{u}_s$, $\partial_r\hat{u}_s$ and $\partial_r\hat{u}_s$ admit the extended far field expansions. Then the extended far field coefficients U_1^1 and U_2^1 determine the coefficients U_j^P for $P \in \{\partial_r, \partial_r^2\}$ and $j \in \{1,2\}$.*

Proof. Integrating the expression

$$\begin{aligned}\frac{\partial}{\partial r}\left[e^{-ikr}\hat{u}_s(r\hat{x},k)\right] &= e^{-ikr}\frac{\partial\hat{u}_s}{\partial r}(r\hat{x},k) - ike^{-ikr}\hat{u}_s(r\hat{x},k) \\ &= \sum_{j=1}^{N} \frac{U_j^{\partial_r}(\hat{x},k) - U_j^1(\hat{x},k)}{r^j} + e^{-ikr}\left[E_N^{\partial_r}(r\hat{x},k) - ikE_N^1(r\hat{x},k)\right],\end{aligned}$$

with respect to r from infinity, we see that

$$\hat{u}_s(r\hat{x},k) = e^{ikr}\sum_{j=1}^{N}\frac{U_j^{\partial_r}(\hat{x},k) - ikU_j^1(\hat{x},k)}{-(j-1)r^{j-1}} + \tilde{E}_N^1(r\hat{x},k),$$

where

$$\tilde{E}_N^1(r\hat{x},k) = e^{ikr}\int_\infty^r \left[E_N^{\partial_r}(s\hat{x},k) - ikE_N^1(s\hat{x},k)\right]e^{-iks}\,ds = \mathcal{O}(k^{N+1})\mathcal{O}(r^{\max\{M_N^1,M_N^{\partial_r}\}+1}).$$

Comparing this with (3.13), we obtain that

$$U_j^{\partial_r}(\hat{x},k) = ikU_j^1(\hat{x},k) - (j-1)U_{j-1}^1(\hat{x},k) \tag{3.27}$$

for $j \in \{1,2\}$, with the interpretation $U_0^1 \equiv 0$.

Analogously integrating twice

$$\frac{\partial^2}{\partial r^2}\left[e^{-ikr}\hat{u}_s(r\hat{x},k)\right] = \sum_{j=1}^{N}\frac{U_j^{\partial_r^2}(\hat{x},k) - 2ikU_j^{\partial_r}(\hat{x},k) - k^2U_j^1(\hat{x},k)}{r^j} + \tilde{\tilde{E}}_N^{\partial_r^2}(r\hat{x},k),$$

where

$$\tilde{\tilde{E}}_N^{\partial_r^2}(r\hat{x},k) = e^{-ikr}\left[E_N^{\partial_r^2}(r\hat{x},k) - 2ikE_N^{\partial_r}(r\hat{x},k) - k^2E_N^1(r\hat{x},k)\right],$$

and substituting (3.27) we get

$$U_j^{\partial_r^2} = -k^2U_j^1 - 2ik(j-1)U_{j-1}^1 + (j-1)(j-2)U_{j-2}^1, \tag{3.28}$$

again interpreting $U_0^1 = U_{-1}^1 = 0$. □

Lemma 3.7. *Let $q \in W_\gamma^{1,\infty}(\mathbb{R}^3,\mathbb{R})$, $\gamma > 21$ be such that the operator $I + \mathcal{G}_0 Q$ is injective on $L^2_{-\delta}$ for some $\delta > 5/2$. Assume that $u_i(x,t) = \int_{S^2} h(x\cdot\omega - t,\omega)\,dS(\omega)$ with $h \in C_0^\infty(\mathbb{R}\times S^2)$. Then the far field pattern $\hat{u}_s^\infty = U_1^1$ determines the extended far field coefficient U_2^1.*

Proof. In spherical coordinates, the equation

$$\left[-\triangle - k^2 + q(r\hat{x})\right]\hat{u}_s(r\hat{x},k) = -q(r\hat{x})\hat{u}_i(r\hat{x},k)$$

reads

$$\left[-\frac{\partial^2}{\partial r^2} - \frac{2}{r}\frac{\partial}{\partial r} - \frac{\triangle_\perp}{r^2} - k^2 + q(r\hat{x})\right]\hat{u}_s(r\hat{x},k) = -q(r\hat{x})\hat{u}_i(r\hat{x},k), \tag{3.29}$$

where $\triangle_\perp$ is the spherical part of the Laplacian. We now plug the extended far field expansions and the relations (3.27) and (3.28) into (3.29). Since

$$|\hat{u}_i(r\hat{x},k)| = \int_{\mathbb{R}}\int_{S^2}|e^{ikt}||h(x\cdot\omega - t,t)|\,dS(\omega)\,dt \le \|h\|_{L^1(\mathbb{R}\times S^2)}$$

is uniformly bounded, and since $\hat{u}_s(r\hat{x}) = \mathcal{O}(1/r)$ as $r \to \infty$ by Lemma 3.3, we see that $q(r\hat{x})\hat{u}_i(r\hat{x},k)$ and $q(r\hat{x})\hat{u}_s(r\hat{x},k)$ are $\mathcal{O}(r^{-\gamma})$. The existence of the extended far field expansions of $\partial_r^2 \hat{u}_s$, $\partial_r \hat{u}_s$ and $\hat{u}_s$ thus implies the existence of a similar expansion for $\triangle_\perp \hat{u}_s$. The calculation

$$\begin{aligned}
&\left| \int_{S^2} \left[\triangle_\perp \hat{u}_s(r\hat{x},k) - \sum_{j=1}^{N} \frac{\triangle_\perp U_j^1(\hat{x},k)}{r^j} \right] \varphi(\hat{x})\, dS(\hat{x}) \right| \\
&\qquad = \left| \int_{S^2} \left[\hat{u}_s(r\hat{x},k) - \sum_{j=1}^{N} \frac{U_j^1(\hat{x},k)}{r^j} \right] \triangle_\perp \varphi(\hat{x})\, dS(\hat{x}) \right| \\
&\qquad \le C w(r)^{M_N^1} w(k)^M \|\triangle_\perp \varphi\|_{L^1(S^2)},
\end{aligned}$$

valid for any $\varphi \in C^\infty(S^2)$, shows that the coefficients of this expansion are $\triangle_\perp U_j^1(\hat{x},k)$. We therefore get

$$\begin{aligned}
&\sum_{j=1}^{L} \left[\frac{-k^2 U_j^1(\hat{x},k) - 2ik(j-1)U_{j-1}^1(\hat{x},k) + (j-1)(j-2)U_{j-2}^1(\hat{x},k)}{r^j} \right. \qquad (3.30) \\
&\qquad \left. + \frac{2ikU_j^1(\hat{x},k) - 2(j-1)U_{j-1}^1(\hat{x},k)}{r^{j+1}} + \frac{\triangle_\perp U_j^1(\hat{x},k)}{r^{j+2}} + \frac{k^2 U_j^1(\hat{x},k)}{r^j} \right] = o\left(\frac{1}{r^L}\right).
\end{aligned}$$

We now use the spherical harmonic expansions

$$U_j^1(\hat{x},k) = \sum_{n=0}^{\infty} \sum_{m=-n}^{n} a_j^{nm}(k) Y_n^m(\hat{x}), \qquad j \in \{1,\dots,L\},$$

where

$$a_j^{nm}(k) = \frac{1}{|S^2|} \int_{S^2} U_j^1(\hat{x},k) \overline{Y_n^m(\hat{x})}\, dS(\hat{x}).$$

Taking the inner product of both sides of (3.30) with $\overline{Y_n^m(\hat{x})}$ and integrating by parts twice, we get

$$\begin{aligned}
&\sum_{j=1}^{L} \left[\frac{-2ik(j-1)a_{j-1}^{nm}(k) + (j-1)(j-2)a_{j-2}^{nm}(k)}{r^j} \right. \\
&\qquad \left. + \frac{2ika_j^{nm}(k) - 2(j-1)a_{j-1}^{nm}(k)}{r^{j+1}} + \frac{n(n+1)a_j^{nm}(k)}{r^{j+2}} \right] = o\left(\frac{1}{r^L}\right).
\end{aligned}$$

This yields the recurrence relations

$$2ik(j-2)a_{j-1}^{nm} = [j-3+n(n+1)]\, a_{j-2}^{nm}, \qquad j \in \{1,\dots,L\}.$$

Our assumption about γ allows us to choose $L = 3$, yielding $a_2^{nm} = \frac{n(n+1)}{2k} i a_1^{nm}$. □

Lemma 3.8. *Let $q \in L^\infty_\gamma(\mathbb{R}^3,\mathbb{R})$, $\gamma > 28$ be such that the operator $I + \mathcal{G}_0 Q$ is injective on $L^2_{-\delta}$ for some $\delta > 5/2$. Let $u_{\rm i}(x,t) = \int_{S^2} h(x\cdot\omega,\omega)\,dS(\omega)$ with $h \in C_0^\infty(\mathbb{R}\times S^2)$. Then the extended far field coefficients U_j^1 and $U_j^{\partial_r}$ determine the coefficients $U_j^{\partial_k^2}$ and $U_j^{\partial_r\partial_k^2}$ for $j \in \{-1,\dots,2\}$.*

Proof. For each $\varphi \in C_0^\infty(S^2\times\mathbb{R})$, we have by (3.11) of Lemma 3.3 that

$$\left|\int_{\mathbb{R}}\int_{S^2}\left[\hat{u}_{\rm s}(r\hat{x},k) - e^{ikr}\sum_{j=1}^{N}\frac{U_j^1(\hat{x},k)}{r^j}\right]\partial_k^2\varphi(\hat{x},k)\,d\hat{x}\,dk\right| \le Cw(r)^{M_N^1}\|\partial_k^2\varphi\|_{L^1_N},$$

with M_N^1 defined by (3.19). Integrating by parts twice, we see that

$$\begin{aligned}
&\left|\int_{\mathbb{R}}\int_{S^2}\left[\partial_k^2\hat{u}_{\rm s}(r\hat{x},k) - e^{ikr}\sum_{j=1}^{N}\frac{-r^2U_j^1(\hat{x},k) + 2ir\partial_k U_j^1(\hat{x},k) + \partial_k^2 U_j^1(\hat{x},k)}{r^j}\right]\varphi(\hat{x},k)\,d\hat{x}\,dk\right| \\
&= \left|\int_{\mathbb{R}}\int_{S^2}\left[\partial_k^2\hat{u}_{\rm s}(r\hat{x},k)\right.\right. \\
&\qquad\left.\left. - e^{ikr}\sum_{j=-1}^{N}\frac{-U_{j+2}^1(\hat{x},k) + 2i\partial_k U_{j+1}^1(\hat{x},k) + \partial_k^2 U_j^1(\hat{x},k)}{r^j}\right]\varphi(\hat{x},k)\,d\hat{x}\,dk\right| \\
&\le Cw(r)^{M_N^1}\|\partial_k^2\varphi\|_{L^1_N}.
\end{aligned}$$

On the other hand, the expansion (3.12) is determined by the estimate

$$\int_{\mathbb{R}}\int_{S^2}\left[\partial_k^2\hat{u}_{\rm s}(r\hat{x},k) - e^{ikr}\sum_{j=-1}^{N}\frac{U_j^{\partial_k^2}(\hat{x},k)}{r^j}\right]\varphi(\hat{x},k)\,d\hat{x}\,dk \le Cw(r)^{M_N^{\partial_k^2}}\|\varphi\|_{L^1_N},$$

with $M_N^{\partial_k^2}$ defined by (3.20). We thus conclude that

$$U_j^{\partial_k^2}(\hat{x},k) = -U_{j+2}^1(\hat{x},k) + 2i\partial_k U_{j+1}^1(\hat{x},k) + \partial_k^2 U_j^1(\hat{x},k).$$

Analogously we see how the coefficients of $\partial_r\hat{u}_{\rm s}$ determine those of $\partial_r\partial_k^2\hat{u}_{\rm s}$. □

Combining the results of these three lemmata, we thus have the following.

Theorem 3.9. *Let $q \in W^{1,\infty}_\gamma(\mathbb{R}^3,\mathbb{R})$ with $\gamma > 28$ be such that the operator $I + \mathcal{G}_0 Q$ is injective on $L^2_{-\delta}$ for some $\delta > 5/2$. Let $u_{\rm i}(x,t) = \int_{S^2} h(x\cdot\omega,\omega)\,dS(\omega)$ with $h \in C_0^\infty(\mathbb{R}\times S^2)$. Then the far field pattern $\hat{u}_{\rm s}^\infty = U_1^1$ determines the extended far field data U_j^P for $j \in \{-1,\dots,2\}$ and $P \in \{1,\partial_k^2,\partial_r,\partial_r\partial_k^2\}$.*

3.5 Scattering amplitude

Scattering data are often formulated in terms of the *scattering amplitude* $A(\hat{x},\omega,k)$. This is defined as the far field of the scattered field corresponding to an incident plane wave: $A(\omega,\hat{x},k) = \hat{v}_{\mathrm{s}}^{\infty}(\hat{x},k;\omega)$, where

$$\left[-\triangle - k^2 + q(x)\right]\hat{v}(x,k;\omega) = 0 \tag{3.31}$$

$$\hat{v}(x,k;\omega) = \hat{v}_{\mathrm{i}}(x,k;\omega) + \hat{v}_{\mathrm{s}}(x,k;\omega) \tag{3.32}$$

$$\hat{v}_{\mathrm{i}}(x,k;\omega) = e^{ikx\cdot\omega} \tag{3.33}$$

$$\lim_{r\to\infty} \frac{\partial \hat{v}_{\mathrm{s}}(r\hat{x},k;\omega)}{\partial r} - ik\hat{v}_{\mathrm{s}}(x,k;\omega) = o\left(\frac{1}{r}\right) \tag{3.34}$$

It is pleasing to see that the scattering amplitude is, indeed, exactly what we need here, too. This is because we want to calculate the inner products of time derivatives of solutions for which

$$u_{\mathrm{i}}(x,t) = \int_{S^2} h_u(x\cdot\omega + t,\omega)\,dS(\omega)$$

with $h_u \in L^2(\mathbb{R}\times S^2)$. In the frequency domain, this is a *Herglotz wave function*, *i.e.*, a wave whose incident part is

$$\hat{u}_{\mathrm{i}}(x,k) = \int_{S^2} e^{ikx\cdot\omega}\hat{h}_u(k,\omega)\,dS(\omega),$$

where $\hat{h}_u(k,\cdot) \in L^2(S^2)$ for almost all $k \in \mathbb{R}$. As could be expected, the scattered part is the corresponding linear combination of scattered plane waves, and analogously for the far field:

Lemma 3.10. *Assume that $q \in W^{1,\infty}_{\mathrm{loc}}(\mathbb{R}^3,\mathbb{R}) \cap L^{\infty}_{\gamma}$ with $\gamma > 13$ and $k \neq 0$. Let*

$$\hat{F}_{\mathrm{i}}(x,k) = \int_{S^2} \hat{h}(\omega)\hat{v}_{\mathrm{i}}(x,k;\omega)\,dS(\omega), \quad \hat{F}_{\mathrm{s}}(x,k) = \int_{S^2} \hat{h}(\omega)\hat{v}_{\mathrm{s}}(x,k;\omega)\,dS(\omega), \tag{3.35}$$

where $\hat{h} \in L^2(S^2)$ and $\hat{v} = \hat{v}_{\mathrm{i}} + \hat{v}_{\mathrm{s}}$ is the scattering solution corresponding to the incident plane wave, i.e., *the solution to (3.31)–(3.34). Then $\hat{F} = \hat{F}_{\mathrm{i}} + \hat{F}_{\mathrm{s}}$ solves the scattering problem*

$$\left[-\triangle - k^2 + q(x)\right]\hat{F}(x,k) = 0 \tag{3.36}$$

$$\frac{\partial \hat{F}_{\mathrm{s}}}{\partial r}(r\hat{x},k) - ik\hat{F}_{\mathrm{s}}(r\hat{x},k) = o\left(\frac{1}{r}\right) \qquad \text{as } r\to\infty, \tag{3.37}$$

the Sommerfeld radiation condition (3.37) holding uniformly over all $\hat{x} \in S^2$. The scattered wave $\hat{F}_{\mathrm{s}}$ has the far field pattern

$$\hat{F}_{\mathrm{s}}^{\infty}(\hat{x},k) = \int_{S^2} \hat{h}(\omega)\hat{v}_{\mathrm{s}}^{\infty}(x,k;\omega)\,dS(\omega),$$

where $\hat{v}_{\mathrm{s}}^{\infty}$ is the far field of $\hat{v}_{\mathrm{s}}$.

Proof. Lemma 2.20 asserts that $\hat{v} = \hat{v}_{\mathrm{i}} + \hat{v}_{\mathrm{s}}$ solves the Lippman-Schwinger equation. Multiplying this equation by $\hat{h}(\omega)$ and integrating over ω, we get

$$\begin{aligned}\hat{F}_{\mathrm{s}}(x,k) &= \int_{S^2} \hat{h}(\omega)\hat{v}_{\mathrm{s}}(x,k;\omega)\,dS(\omega) \\ &= -\int_{S^2}\int_{\mathbb{R}^3} \Phi(x-y)q(y)\hat{h}(\omega)\hat{v}(y,k;\omega)\,dy\,dS(\omega) \\ &= -\int_{\mathbb{R}^3} \Phi(x-y)q(y)\hat{F}(y,k)\,dy.\end{aligned} \tag{3.38}$$

Here we can change the order of integration since Lemmata 2.4 and 2.9 show that for any δ between $5/2$ and $(\gamma-3)/2$,

$$\|\hat{v}(\cdot,k;\omega)\|_{L^2_{-\delta}} \le \|(I+\mathcal{G}_kQ)^{-1}\mathcal{G}_kQ + I\|_{\mathcal{L}(L^2_{-\delta})}\|\hat{v}_{\mathrm{i}}(\cdot,k;\omega)\|_{L^2_{-\delta}}$$

is uniformly bounded with respect to $\omega \in S^2$, and thus both

$$\begin{aligned}&\int_{S^2}\int_{B(x,1)} |\Phi(x-y)||q(y)||\hat{h}(\omega)||\hat{v}(y,k;\omega)|\,dy\,dS(\omega) \\ &\qquad \le C\int_{S^2} |\hat{h}(\omega)| \left\| \frac{1}{|x-\cdot|}\right\|_{L^2(B(x,1))} \|q\hat{v}(\cdot,k;\omega)\|_{L^2(B(x,1))}\,dS(\omega) \\ &\qquad \le C\int_{S^2} |\hat{h}(\omega)|\|q\|_{L^\infty_\delta}\|\hat{v}(\cdot,k;\omega)\|_{L^2_{-\delta}}\,dS(\omega)\end{aligned}$$

and

$$\begin{aligned}&\int_{S^2}\int_{\mathbb{R}^3\setminus B(x,1)} |\Phi(x-y)||q(y)||\hat{h}(\omega)||\hat{v}(y,k;\omega)|\,dy\,dS(\omega) \\ &\qquad \le C\int_{S^2} |\hat{h}(\omega)|\|q\|_{L^2_\delta}\|\hat{v}(\cdot,k;\omega)\|_{L^2_{-\delta}}\,dS(\omega)\end{aligned} \tag{3.39}$$

are finite.

Again by Lemma 2.20, the Lippman-Schwinger equation (3.38) implies that $\hat{F}$ solves the scattering problem. The far field pattern is

$$\begin{aligned}\hat{F}_{\mathrm{s}}^\infty(\hat{x},k) &= -\int_{\mathbb{R}^3} e^{-ik\hat{x}\cdot y}q(y)\hat{F}(y)\,dy \\ &= -\int_{\mathbb{R}^3}\int_{S^2} e^{-ik\hat{x}\cdot y}q(y)\hat{h}(\omega)\hat{v}(y,k;\omega)\,dS(\omega)\,dy \\ &= \int_{S^2} \hat{h}(\omega)\hat{v}_{\mathrm{s}}^\infty(\hat{x},k;\omega)\,dS(\omega)\,dy,\end{aligned}$$

changing the order of integration being permitted by a calculation almost identical to (3.39). □

Our variant of the Blagoveščenskiĭ identity thus reads:

Theorem 3.11. *Let $q \in W^{1,\infty}_{\gamma}(\mathbb{R}^3,\mathbb{R})$ with $\gamma > 28$ be such that the operator $I + \mathcal{G}_0 Q$ is injective on $L^2_{-\delta}$ for some $\delta > 5/2$. Assume that the incoming incident waves*

$$u_{\mathrm{i}}(x,t) = \int_{S^2} h_u(x\cdot\omega - t,\omega)\, dS(\omega) \quad \text{and} \quad v_{\mathrm{i}}(x,t) = \int_{S^2} h_v(x\cdot\omega - t,\omega)\, dS(\omega)$$

are known, with $h_u, h_v \in C_0^\infty(\mathbb{R}\times S^2)$, as well as the scattering amplitude. Then the inner product

$$\int_{\mathbb{R}^3} u(x,s)\overline{v(x,t)}\, dx$$

can be calculated for all $s,t \in \mathbb{R}$ for which $u(\cdot,s)$ and $v(\cdot,t)$ are square integrable.

3.6 Inner products of time derivatives

Theorem 3.11 may not be very useful if we cannot be sure that $u(\cdot,t)$ is square integrable. However, we can work with the time derivative, which is always known to be square integrable for compactly supported potentials by Lemma 2.19.

It actually follows that if $u_{\mathrm{i}} \in \tilde{H}$ and $u_{\mathrm{i}}(\cdot,t_0)$, $u_{\mathrm{s}}(\cdot,t_0)$ or $u(\cdot,t_0)$ is square integrable for some time $t_0 \in \mathbb{R}$, then it is for all other times, too, since

$$\|u(\cdot,t)\|_{L^2} \le \|u(\cdot,t_0)\|_{L^2} + \Big|\int_{t_0}^{t} \|\partial_t u(\cdot,s)\|_{L^2}\, ds\Big| \le \|u(\cdot,t_0)\|_{L^2} + |t-t_0|\|h\|_{L^2(\mathbb{R}\times S^2)}$$

and analogously for u_{i} and u_{s}. The assumption of square integrability at any one time would thus be sufficient for the inner products in Theorem 3.11 to make sense. However, we can avoid making this additional assumption by calculating directly the inner products of time derivatives of solutions instead. Here we assume that the potential has compact support in order to be able to use Lemma 2.19; when this result is applied in Section 5, this assumption will also be used for other purposes.

Theorem 3.12. *Let $q \in W^{1,\infty}_{\mathrm{comp}}(\mathbb{R}^3,\mathbb{R})$ be such that the operator $I + \mathcal{G}_0 Q$ is injective on $L^2_{-\delta}$ for some $\delta > 5/2$. Then if the scattering amplitude is known, the inner product*

$$\int_{\mathbb{R}^3} \partial_s u(x,s)\overline{\partial_t v(x,t)}\, dx \tag{3.40}$$

can be calculated for all known incident waves $u_{\mathrm{i}}, v_{\mathrm{i}} \in \tilde{H}$ and all $s,t \in \mathbb{R}$.

Proof. Let h_u and $h_v \in L^2(\mathbb{R}\times S^2)$ be the translation representations of the incident waves u_{i} and v_{i}, respectively. Choose sequences of functions $h_u^j, h_v^j \in C_0^\infty(\mathbb{R}\times S^2)$ converging in $L^2(\mathbb{R}\times S^2)$ to h_u and h_v, respectively. Set

$$u_{\mathrm{i}}^j(x,t) = \int_{S^2} h_u^j(x\cdot\omega - t,\omega)\, dS(\omega) \quad \text{and} \quad v_{\mathrm{i}}^j(x,t) = \int_{S^2} h_v^j(x\cdot\omega - t,\omega)\, dS(\omega)$$

and let $u^j = u^j_{\rm i} + u^j_{\rm s}$ and $v^j = v^j_{\rm i} + v^j_{\rm s}$ be the corresponding total waves. Now $\partial_s u(\cdot, s)$ and $\partial_t v(\cdot, t)$ are solutions of the plasma wave equation (1.1) corresponding to incident waves

$$\partial_s u^j_{\rm i}(x,s) = \int_{S^2} \partial_s h^j_u(x\cdot\omega - s, \omega)\, dS(\omega) \quad \text{and} \quad \partial_t v^j_{\rm i}(x,t) = \int_{S^2} \partial_t h^j_v(x\cdot\omega - t, \omega)\, dS(\omega).$$

Since these incident waves satisfy the assumptions of Theorem 3.1, the inner products

$$\int_{\mathbb{R}^3} \partial_s u^j(x,s)\overline{\partial_t v^j(x,t)}\, dx$$

can be calculated; the s and t derivatives translate into a multiplication of the integrand in (3.8) by $-\sigma\tau$, which only requires one more derivative for h^j_u and h^j_v, which were assumed smooth anyway. Since

$$\begin{aligned}\|\partial_s u(\cdot,x) - \partial_s u^j(\cdot,s)\|_{L^2} &\le \tilde{E}_0(u - u^j, s)\\ &\le C\tilde{E}_q(u - u^j, s)\\ &= C\tilde{E}_0(u_{\rm i} - u^j_{\rm i}, 0)\\ &= C\|h_u - h^j_u\|_{L^2(\mathbb{R}\times S^2)} \to 0\end{aligned}$$

and analogously for $\partial_t v - \partial_t v^j$, these inner products converge to the inner product (3.40). □

4 Scattering control

We shall now take the next step towards solving the inverse problem of the plasma wave equation (1.1) using scattering data. The previous section showed how inner products of solutions can be calculated from these data. Now we show how to select the solutions in such a way that knowing their inner products allows us to solve the inverse problem. The potential q will be assumed to be compactly supported in this section.

Our strategy will be the following: By Rellich's lemma [CK98, Thm. 2.13], the scattering data determine the solutions corresponding to incident plane waves at points outside the support of the potential. The mixed reciprocity relation [Pot01, Thm. 2.2.4] shows that these are the same as far field data corresponding to point sources outside the support of the potential. We show that using this information, we can construct a superposition of plane waves which gives, from a certain point in time on, the same solution as that induced by a point source.

Then we show that, with point sources in a small set outside the support of the potential, it is possible to control the solutions in the support at any particular time. In particular, we can excite a wave that is nonzero in a neighbourhood of any point. In Section 5, we calculate inner products of such waves using the Blagoveščenskiĭ identity, and choosing the waves in an appropriate way, we eventually determine the values of the waves at any point. The potential is then solved from (1.1).

The method used here is a variant of the *boundary control (BC) method*, pioneered by Belishev and Kurylev [Bel90, BK92b, Bel97]. As the name suggests, the BC method traditionally deals with a boundary value problem. There the control property says that by placing sources on a piece of the boundary from time 0 to time T, it is possible to approximately control at time T the domain of dependence, which is the set of points that the waves have reached in this time. Now the sources on a boundary are replaced by the simulated sources described above.

4.1 Simulating point sources

In this section, we investigate the possibility of simulating point sources by sending in plane waves. In other words, we would like to express the causal Green's function $g(x,t;x_0)$, *i.e.*, the solution of the inhomogeneous plasma wave equation with a point source,

$$[\partial_t^2 - \triangle_x + q(x)]g(x,t;x_0) = \delta_{x_0}(x)\delta_0(t) \tag{4.1}$$

$$g|_{t<0} = 0, \tag{4.2}$$

as a superposition of scattering solutions $v = v_\mathrm{i} + v_\mathrm{s}$ corresponding to incident plane waves,[8]

$$\begin{aligned}[\partial_t^2 - \triangle_x + q(x)]v(x,t;\omega) &= 0 \\ v_\mathrm{i}(x,t;\omega) &= \delta(t - x\cdot\omega) \\ v(x,t;\omega) &= 0 \quad \text{when } t < x\cdot\omega.\end{aligned}$$

Of course, this is not possible, since plane waves satisfy the homogeneous plasma wave equation (1.1), so that no superposition of them can give the delta source in (4.1).

However, what we ultimately want to do is to control solutions $u(\cdot,T)$ of the plasma wave equation by sending in superpositions of plane waves. This is done in two steps, which we shall first describe unrigorously, and shortly thereafter proceed to justify the formal calculations. The first step is to express $u(\cdot,T)$ as a linear combination of Green's functions,

$$u(x,t) = \int_A \int_0^T H(y,s)g(x,t-s;y)\,ds\,dy =: u^H(x,t),$$

where A is a bounded set in the complement of $\operatorname{supp} q$; this question of what we call interior control will be considered in Section 4.2 below. In the second step, we try to send in a superposition of plane waves that has the same effect at time T as the above linear combination of Green's functions.

The second step can be achieved as follows using the odd continuation of the causal Green's function: Set

$$f(x,t;x_0) := g(x,t;x_0) - g(x,-t;x_0), \tag{4.3}$$

[8] The scattered wave u_s is sometimes called a *retarded pulse*. [New85]

where $g(x,t;x_0)$ is the causal Green's function with source at x_0, *i.e.*, the solution of (4.1)–(4.2). This function gives the same effect as the causal Green's function: since $f(x,t;x_0) = g(x,t;x_0)$ for $t \geq 0$,

$$\int_A \int_0^T H(y,s) f(x,t-s;y)\,ds\,dy = \int_A \int_0^T H(y,s) g(x,t-s;y)\,ds\,dy = u^H(x,t)$$

when $t \geq T$. In contrast to the Green's function g, the function f satisfies

$$\left[\partial_t^2 - \triangle + q(x)\right] f(x,t;x_0) = \delta_{x_0}(x)\delta_0(t) - \delta_{x_0}(x)\delta_0(t) = 0$$

and it turns out that it can, indeed, be expressed as a formal linear combination of plane waves,[9]

$$f(x,t;x_0) = F(x,t;x_0) := \int_{S^2} \int_{-\infty}^{\infty} m(\omega,s;x_0) v(x,t-s;\omega)\,ds\,dS(\omega), \tag{4.5}$$

if the density m is chosen appropriately. We shall now continue these formal calculations and derive a necessary condition for m, and then proceed to show that this condition is also sufficient for (4.5) to hold.

In analogy with Theorem 2.24, the Fourier transform gives $\hat{f}(x,k;x_0) = \hat{g}(x,k;x_0) - \overline{\hat{g}(x,k;x_0)}$, where

$$\left[-\triangle_x - k^2 + q(x)\right]\hat{g}(x,k;x_0) = \delta_{x_0}(x) \tag{4.6}$$

$$\frac{\partial \hat{g}}{\partial r}(r\hat{x},k;x_0) - ik\hat{g}(r\hat{x},k;x_0) = o\left(\frac{1}{r}\right) \quad \text{as } r \to \infty, \tag{4.7}$$

the Sommerfeld radiation condition (4.7) holding uniformly over all directions $\hat{x} \in S^2$. Thus, $\hat{f}$ satisfies the frequency domain plasma wave equation:

$$\left[-\triangle - k^2 + q(x)\right]\hat{f}(x,k;x_0) = 0. \tag{4.8}$$

On the other hand, $\hat{F}$ is a scattered Herglotz wave, *i.e.*,

$$\hat{F}(x,k;x_0) = \int_{S^2} \hat{m}(\omega,k;x_0)\hat{v}(x,k;\omega)\,dS(\omega), \tag{4.9}$$

where $\hat{v} = \hat{v}_\mathrm{i} + \hat{v}_\mathrm{s}$ is the scattered wave corresponding to an incident plane wave defined by (3.31)–(3.34). Lemma 3.10 tells us that $\hat{F} = \hat{F}_\mathrm{i} + \hat{F}_\mathrm{s}$, where $\hat{F}_\mathrm{i}$ is the corresponding superposition of incident plane waves, and $\hat{F}_\mathrm{s}$ satisfies the Sommerfeld radiation condition,

$$\frac{\partial \hat{F}_\mathrm{s}}{\partial r}(r\hat{x},k;x_0) - ik\hat{F}_\mathrm{s}(r\hat{x},k;x_0) = o\left(\frac{1}{r}\right) \qquad \text{as } r \to \infty, \tag{4.10}$$

[9] Up to a reflection in time and a permutation of the variables, m is the translation representation of the incident part

$$F_\mathrm{i}(x,t;x_0) = \int_{S^2} \int_{\mathbb{R}} m(\omega,s;x_0)\delta(t-s-x\cdot\omega)\,ds\,dS(\omega) = \int_{S^2} m(\omega,t-x\cdot\omega;x_0)\,dS(\omega). \tag{4.4}$$

uniformly in all directions $\hat{x} \in S^2$.

The incident part has the following far field (cf. [Mel94, GY99]):

Lemma 4.1. *Let*

$$\hat{F}_{\mathrm{i}}(x,k;x_0) = \int_{S^2} \hat{m}(\omega,k;x_0)\hat{v}_{\mathrm{i}}(x,k;\omega)\,dS(\omega).$$

with $\hat{m}(k,\cdot;x_0) \in C^4(S^2)$. *Fix* $k \in \mathbb{R}$ *and* $x_0 \in \mathbb{R}^3$. *Then*

$$\hat{F}_{\mathrm{i}}(x,k;x_0) = \frac{e^{ik|x|}}{|x|}\frac{2\pi i}{k}\hat{m}(\hat{x},k;x_0) - \frac{e^{-ik|x|}}{|x|}\frac{2\pi i}{k}\hat{m}(-\hat{x},k;x_0) + \mathcal{O}\left(\frac{1}{|x|^2}\right)$$

as $|x| \to \infty$.

Proof. As k and x_0 are fixed, write $\hat{m}(\omega) = \hat{m}(k,\omega;x_0)$ to simplify notation. Use the method of stationary phase [Hör90, Thm. 7.7.5] as follows: Fix $\hat{x} \in S^2$. Using two smooth coordinate charts $(U_\pm, \varphi_\pm)$ covering S^2, with $\varphi_\pm(\pm\hat{x}) = 0$, and an associated partition of unity $\{\chi_+, \chi_-\}$, we get

$$\hat{F}_{\mathrm{i}}(x,k;x_0) = \int_{S^2} e^{ikr\hat{x}\cdot\omega}\hat{m}(\omega)\,dS(\omega) = \int_{\varphi_+U_+} e^{irf_+(y)}\hat{m}_+(y)\,dy + \int_{\varphi_-U_-} e^{irf_-(y)}\hat{m}_-(y)\,dy,$$

where

$$f_\pm(y) = k\varphi_\pm^{-1}(y)\cdot\hat{x} \qquad \hat{m}_\pm(y) = \chi_\pm(\varphi_\pm^{-1}(y))\hat{m}(\varphi_\pm^{-1}(y))J_{\varphi_\pm^{-1}}(y).$$

Here $\operatorname{Im} f_\pm \equiv 0$, $\nabla f_\pm(0) = 0$, $\det(\partial_j\partial_m f_\pm(0)) \neq 0$ and $\nabla f_\pm(y) \neq 0$ for all $y \in \varphi_\pm U_\pm \setminus \{0\}$. Thus the first order stationary phase expansion gives

$$\int_{\varphi_\pm U_\pm} e^{irf_\pm(y)}\hat{m}_\pm(y)\,dy = c_\pm e^{irf_\pm(0)}\frac{\hat{m}_\pm(0)}{r} + \mathcal{O}\left(\frac{1}{r^2}\right).$$

with $c_\pm$ independent of $\hat{m}$. Thus,

$$\hat{F}_{\mathrm{i}}(x,k;x_0) = \frac{e^{ik|x|}}{|x|}c_+\hat{m}(\hat{x}) + \frac{e^{-ik|x|}}{|x|}c_-\hat{m}(-\hat{x}) + \mathcal{O}\left(\frac{1}{|x|^2}\right).$$

Choosing φ to be, for instance, the stereographic projection, we see that $c_\pm = \pm 2\pi i/k$. □

Combining Lemmata 3.10 and 4.1 shows that if $\hat{m}(k,\cdot;x_0) \in C^4(S^2)$, the full Herglotz wave $\hat{f}$ has the asymptotic behaviour

$$\int_{S^2} \hat{m}(\omega)\hat{v}(x,k;\omega)\,dS(\omega) = \frac{e^{ik|x|}}{|x|}\left[\frac{2\pi i}{k}\hat{m}(\hat{x},k;x_0) + \int_{S^2}\hat{m}(\omega)\hat{v}_{\mathrm{s}}^{\infty}(\hat{x},k;\omega)\,dS\right] + - \frac{e^{-ik|x|}}{|x|}\frac{2\pi i}{k}\hat{m}(-\hat{x},k;x_0) + \mathcal{O}\left(\frac{1}{|x|^2}\right). \quad (4.11)$$

Now since $\hat{g}$ satisfies the Helmholtz equation outside $\operatorname{supp} q$ and the Sommerfeld radiation condition, it has the far field expansion [CK98, Thm. 2.5]

$$\hat{g}(x,k;x_0) = \frac{e^{ik|x|}}{|x|}\hat{g}^\infty(\hat{x},k;x_0) + o\left(\frac{1}{|x|}\right)$$

and thus

$$\hat{f}(x,k;x_0) = \frac{e^{ik|x|}}{|x|}\hat{g}^\infty(\hat{x},k;x_0) - \frac{e^{-ik|x|}}{|x|}\overline{\hat{g}^\infty(\hat{x},k;x_0)} + o\left(\frac{1}{|x|}\right). \tag{4.12}$$

In order for $\hat{f}$ to equal $\hat{F}$, the coefficients of the $e^{\pm ik|x|}/|x|$ terms must match in (4.11) and (4.12). In particular, to fulfill this requirement for the antiradiating part, *i.e.*, the second term on the right hand side of (4.12), we must choose $\hat{m}(\omega,k;x_0) = k\overline{\hat{g}^\infty(-\omega,k;x_0)}/2\pi i = k\hat{g}^\infty(-\omega,-k;x_0)/2\pi i$. Actually this is a consistent choice, as the radiating parts match automatically:

Theorem 4.2. *Assume that $q \in C_0^1(\mathbb{R}^3,\mathbb{R})$. Let $\hat{f}(x,k;x_0) = \hat{g}(x,k;x_0) - \overline{\hat{g}(x,k;x_0)}$, where $\hat{g}$ is a solution to (4.6)–(4.7). Set*

$$\hat{m}(\omega,k;x_0) = \frac{k}{2\pi i}\hat{g}^\infty(-\omega,-k;x_0).$$

Then

$$\hat{f}(x,k;x_0) = \int_{S^2}\hat{m}(\omega,k;x_0)\hat{v}(x,k;\omega)\,dS(\omega). \tag{4.13}$$

If $x_0 \notin \operatorname{supp} q$, the scattering amplitude determines $\hat{m}$.

Proof. As before, let

$$\hat{F}_{\mathrm{i}}(x,k;x_0) = \int_{S^2}\hat{m}(\omega,k;x_0)\hat{v}_{\mathrm{i}}(x,k;\omega)\,dS(\omega)$$
$$\hat{F}_{\mathrm{s}}(x,k;x_0) = \int_{S^2}\hat{m}(\omega,k;x_0)\hat{v}_{\mathrm{s}}(x,k;\omega)\,dS(\omega)$$

and $\hat{F} = \hat{F}_{\mathrm{i}} + \hat{F}_{\mathrm{s}}$. Observe that Lemma 4.1 can be applied, since $\hat{g}^\infty(\cdot,-k;x_0)$ is analytic on the unit sphere, and in particular four times continuously differentiable.

Fix $k \in \mathbb{R}$ and write $\varphi = \hat{F} - \hat{f}$. By (4.8) and Lemma 3.10,

$$\left[-\triangle - k^2 + q(x)\right]\varphi(x) = 0.$$

The function φ also satisfies the Sommerfeld radiation condition: Differentiating under the integral and using Lemma 4.1 we see that

$$\begin{aligned}\frac{\partial\hat{F}_{\mathrm{i}}}{\partial r}(x,k;x_0) &= ik\int_{S^2}\hat{x}\cdot\omega\,\hat{m}(\omega,k;x_0)\hat{v}_{\mathrm{i}}(x,k;\omega)\,dS(\omega)\\ &= -\frac{e^{ikr}}{r}2\pi\hat{m}(\hat{x},k;x_0) - \frac{e^{-ikr}}{r}2\pi\hat{m}(-\hat{x},k;x_0) + \mathcal{O}\left(\frac{1}{|x|^2}\right).\end{aligned}$$

Combining this with (4.7) and (4.10), and applying Lemma 4.1 again, we see that

$$\frac{\partial\varphi}{\partial r}(x) - ik\varphi(x) = -\frac{e^{-ikr}}{r}4\pi\hat{m}(-\hat{x},k;x_0) - \frac{e^{-ikr}}{r}2ik\overline{\hat{g}^{\infty}(\hat{x},k;x_0)} + o\left(\frac{1}{r}\right) = o\left(\frac{1}{r}\right)$$

by assumption, uniformly in all directions $\hat{x} \in S^2$. Since the frequency domain scattering problem is uniquely solvable [CK98, Thm. 8.7], we deduce that $\hat{\varphi} \equiv 0$.

The final statement follows from the mixed reciprocity relation [Pot01, Thm. 2.2.4]:

$$\hat{g}^{\infty}(\omega,k;x_0) = \frac{1}{4\pi}\hat{v}_s(x_0,k;-\omega), \tag{4.14}$$

and $\hat{v}_s(x_0,k;-\omega)$ is determined by $\hat{v}_s^{\infty}(\cdot,k;\omega) = A(\cdot,\omega,k)$ by Rellich's lemma [CK98, Thm. 2.13]. □

Remark. It is interesting to note that substituting (4.14) into (4.13) gives

$$\hat{f}_s(x,k;x_0) = \frac{1}{4\pi}\int_{S^2}\overline{\hat{v}_s(x_0,k;\omega)}\hat{v}_s(x,k;\omega)\,dS(\omega),$$

i.e., the scattered solution with a simulated point source is given by the correlation of the values at x and x_0 of scattered plane waves.

We now record the observation that (4.6)–(4.7) indeed has a unique solution.

Lemma 4.3. *Let $q \in C_0^1(\mathbb{R}^3,\mathbb{R})$. Then for all $k \neq 0$, (4.6)–(4.7) has a unique solution*

$$\hat{g}(\cdot,k;x_0) = (I + \mathcal{G}_k Q)^{-1}\Phi(\cdot - x_0,k).$$

Proof. Since both $\hat{g}$ and Φ satisfy the Sommerfeld radiation condition, we see as in the proof of Lemma 2.20 that (4.6)–(4.7) is equivalent with the integral equation

$$\hat{g}(\cdot,k;\varphi) - \Phi(\cdot - x_0,k) = -\mathcal{G}_k Q\hat{g}(\cdot,k;\varphi), \tag{4.15}$$

which has a unique solution by Lemmata 2.9 and 2.10. □

In the time domain analogue (4.5) of the superposition formula (4.13), the product of the two functions of k translates into a convolution in t. To make this rigorous, this convolution of two bounded functions could be interpreted as being defined as the inverse Fourier transform of (4.13). However, if the sources are smoothed in time, the convolution can be viewed in the context of distributions and L^p functions. To this end, we shall use the following two lemmata:

Lemma 4.4. *Let $K \subset \mathbb{R}^3$ be compact and assume that the $q \in C_0^1(\mathbb{R}^3,\mathbb{R})$ is such that the operator $I + \mathcal{G}_0 Q$ is injective on $L^2_{-\delta}$ for some $\delta > 5/2$. Then there is a constant $C > 0$, only depending on q and K, such that*

$$\left|\partial_k^l \hat{g}^{\infty}(\omega,k;x_0)\right| \leq C$$

for all $l \in \{0,1,2\}$, $\omega \in S^2$, $k \in \mathbb{R}\setminus\{0\}$ and $x_0 \in K$.

Proof. Inserting the expansion (3.9) with $N = 1$ into (4.15) gives

$$\hat{g}(\hat{x}, k; x_0) = \frac{e^{ik|x|}}{4\pi|x|}\left[e^{-ik\hat{x}\cdot x_0} - \int_{\mathbb{R}^3} e^{-ik\hat{x}\cdot y} q(y)\hat{g}(y, k; x_0)\, dy\right] + o\left(\frac{1}{|x|}\right)$$

as $|x| \to \infty$, *i.e.*,

$$\partial_k^l \hat{g}^\infty(\omega, k; x_0) = \tfrac{(-i\omega\cdot x_0)^l}{4\pi} e^{-ik\omega\cdot x_0} - \sum_{j=0}^{l} \tbinom{j}{l} \int_{\mathbb{R}^3} \tfrac{(-ik\omega\cdot y)^{l-j}}{4\pi} e^{-ik\omega\cdot y} q(y)\partial_k^j \hat{g}(y, k; x_0)\, dy.$$

Therefore,

$$\begin{aligned}
\left|\partial_k^l \hat{g}^\infty(\omega, k; x_0)\right| &\le C\left[1 + \sum_{j=0}^{l} \int_{\mathbb{R}^3} |q(y)|\left|\partial_k^j \hat{g}(y, k; x_0)\right| dy\right] \\
&\le C\left[1 + \|q\|_{L^2_\delta} \|\partial_k^j \hat{g}(\cdot, k; x_0)\|_{L^2_{-\delta}}\right].
\end{aligned}$$

This is bounded by Corollary 2.13 since

$$\partial_k^j \hat{g}(\cdot, k; x_0) = \sum_{m=0}^{j} \binom{m}{j} \left[\partial_k^{j-m}(I + \mathcal{G}_k Q)^{-1}\right] \partial_k^m \Phi(\cdot - x_0, k)$$

and $\partial_k^m \Phi(\cdot - x_0, k)$ is bounded in $L^2_{-\delta}$ uniformly for all $x_0 \in K$ and $k \in \mathbb{R}$. □

Lemma 4.5. *Let $\hat{m}$ be as in Theorem 4.2. Let $q \in C_0^1(\mathbb{R}^3, \mathbb{R})$ be such that the operator $I + \mathcal{G}_0 Q$ is injective on $L^2_{-\delta}$ for some $\delta > 5/2$. Fix $\eta \in C_0^\infty(\mathbb{R})$ and a compact set $K \subset \mathbb{R}^3$. Then there is a constant C, depending on q, η and K, such that*

$$|(\eta * m)(\omega, t; x_0)| = \left|\mathcal{F}^{-1}\left[\hat{\eta}(k)\hat{m}(\omega, k; x_0)\right](t)\right| \le C w(t)^{-2}$$

for all $\omega \in S^2$, $t \in \mathbb{R}$ and $x_0 \in K$.

Proof. For all $l \in \{0, 1, 2\}$,

$$\partial_k^l \left(\hat{\eta}(k)\hat{m}(\omega, k; x_0)\right) = \frac{k}{2\pi i} \sum_{j=0}^{l} \binom{j}{l} \partial_k^{l-j} \hat{\eta}(k) \partial_k^j \hat{g}^\infty(-\omega, -k; x_0) = \mathcal{O}\left(|k|^{-n}\right) \tag{4.16}$$

for all $n \in \mathbb{N}$ by Lemma 4.4 and the fact that $\hat{\eta} \in \mathcal{S}$. The convolution theorem for $\eta \in C_0^\infty \subset \mathcal{E}'$ and $m \in L^\infty \subset \mathcal{S}'$ shows that $\eta * m = \mathcal{F}^{-1}(\hat{\eta}\hat{m})$ and thus

$$\begin{aligned}
\left\|t^l \left[\eta * m(\omega, \cdot; x_0)\right](t)\right\|_{L^\infty(t)} &\le \left\|\partial_k^l \left[\hat{\eta}(k)\hat{m}(\omega, k; x_0)\right]\right\|_{L^1(k)} \\
&\le \left\|w^2(k)\partial_k^l \left[\hat{\eta}(k)\hat{m}(\omega, k; x_0)\right]\right\|_{L^\infty(k)} \|w^{-2}\|_{L^1},
\end{aligned}$$

whose boundedness is asserted by (4.16). □

In the derivation of the condition $\hat{m}(\omega,k;x_0) = k\hat{g}^\infty(-\omega,-k;x_0)/2\pi i$, we used the Sommerfeld radiation condition (4.7) to get a causal fundamental solution in the time domain. In the proof of this implication, we shall use the following estimate:

Lemma 4.6. *There is a constant $C > 0$ such that*

$$\|\mathcal{G}_z\|_{\mathcal{L}(L^2_\delta, L^2_{-\delta})} \leq \frac{C}{|z|}$$

for all $z \in \overline{\mathbb{C}_+} \setminus \{0\}$.

Proof. The proof follows the lines of [Päi04, Thm. 3.1], where

$$(\mathcal{G}_z f, g)_{L^2} = (\mathcal{F}(\mathcal{G}_z f), \hat{g})_{L^2} = \int_{\mathbb{R}^3} \frac{\hat{f}(\xi)\hat{g}(\xi)}{|\xi|^2 - z^2} d\xi$$

is estimated. As z is complex, some of the details are slightly more tedious when estimating the integrand in the subdomains $||\xi| - |z|| < |z|/2$ and $||\xi| - |z|| \geq |z|/2$. □

Using this estimate, we can now prove that the radiating fundamental solution indeed satisfies the causal support condition:

Lemma 4.7. *Let $q \in C^1_0(\mathbb{R}^3, \mathbb{R})$ be such that the operator $I + \mathcal{G}_0 Q$ is injective on $L^2_{-\delta}$ for some $\delta > 5/2$. Then the inverse Fourier transform g of the solution of (4.6)–(4.7) is the unique causal Green's function,* i.e., *the unique solution of (4.1)–(4.2).*

Proof. The Fourier transform of (4.6) immediately gives (4.1). To see the causal support condition, it suffices to show that the function

$$\mathbb{R} \ni t \mapsto \langle g(\cdot, t; x_0), \varphi \rangle =: G(t) \in \mathbb{C}$$

is supported in $[0, \infty)$ for any test function $\varphi \in C^\infty_0(\mathbb{R}^3)$. We shall do this using the Paley-Wiener theorem.

Fix $\varphi \in C^\infty_0(\mathbb{R}^3)$. For all $\psi \in C^\infty_0(\mathbb{R})$,

$$\begin{aligned} \langle \hat{G}, \psi \rangle &= \langle \langle g(\cdot, t; x_0), \varphi \rangle, \widehat{\psi}(t) \rangle_t \\ &= \langle g(\cdot, \cdot; x_0), \varphi \otimes \widehat{\psi} \rangle \\ &= \langle \hat{g}(\cdot, \cdot; x_0), \varphi \otimes \psi \rangle \\ &= \langle \langle \hat{g}(\cdot, k; x_0), \varphi \rangle, \psi(k) \rangle_k. \end{aligned}$$

By Lemma 4.3 and the limiting absorption principle, $\hat{G}(k) = \lim_{\varepsilon \searrow 0} \hat{G}(k + i\varepsilon)$, where

$$\hat{G}(z) = \langle (I + \mathcal{G}_z Q)^{-1} \Phi(\cdot - x_0, z), \varphi \rangle .$$

Estimate

$$|\hat{G}(z)| |\langle (I + \mathcal{G}_z Q)^{-1} \Phi(\cdot - x_0, z), \varphi \rangle| \leq \|(I + \mathcal{G}_z Q)^{-1}\|_{\mathcal{L}(L^2_{-\delta})} \|\Phi(\cdot - x_0, z)\|_{L^2_{-\delta}} \|\varphi\|_{L^2_\delta}.$$

Now use Lemma 4.6:

$$\left\|(I+\mathcal{G}_z Q)^{-1}\right\|_{\mathcal{L}(L^2_{-\delta})} \leq \frac{1}{1-\frac{C}{|z|}} \leq 2$$

when $|z|>2C$, with C the constant of Lemma 4.6. Therefore $\hat{G}(z)$ is bounded in the set $\{z\in\mathbb{C}_+ \mid |z|>2C\}$. In the compact set $\{z\in\mathbb{C}_+ \mid |z|\leq 2C\}$, $\hat{G}(z)$ is bounded by continuity.

Write $K=[0,\infty)$. Its support function is

$$H_K(\xi)=\sup_{x\in K} x\cdot\xi = \begin{cases}\infty & \text{when } \xi>0\\ 0 & \text{when } \xi\leq 0.\end{cases}$$

The boundedness of $\hat{G}$ in the upper half plane can be written as[10]

$$|\hat{G}(z)|\leq C(1+|z|)^0 e^{H_K(\mathrm{Im}(-z)-0)} \qquad \text{when } H_K(\mathrm{Im}(-z)-0)<\infty.$$

The Paley-Wiener theorem [Hör90, Thm. 7.4.3] thus implies that $\operatorname{supp} G\subset K=[0,\infty)$.

To see uniqueness, let $\tilde{g}$ be another solution of (4.1)–(4.2). Then $u:=g-\tilde{g}$ satisfies

$$\left[\partial_t^2-\triangle+q(x)\right]u=0$$
$$u|_{t<0}=0,$$

which implies that $u=0$. □

We are now ready to write the time domain result concerning simulated sources smoothed in time.

Theorem 4.8. *Assume that $q\in C_0^1(\mathbb{R}^3,\mathbb{R})$ is such that the operator $I+\mathcal{G}_0 Q$ is injective on $L^2_{-\delta}$ for some $\delta>5/2$. Let*

$$\hat{f}_\eta(x,k;x_0)=\hat{\eta}(k)\int_{S^2}\hat{m}(\omega,k;x_0)\hat{v}(x,k;\omega)\,dS(\omega)$$

where $\eta\in C_0^\infty((-\varepsilon,\varepsilon))$,

$$\hat{m}(\omega,k;x_0)=\frac{k}{2\pi i}\hat{g}^\infty(-\omega,-k;x_0)$$

and g is the causal Green's function of the plasma wave equation, i.e., *the solution to (4.1)–(4.2). Then $f_\eta:=\mathcal{F}^{-1}\hat{f}_\eta=f(x,\cdot;x_0)*\eta$ satisfies*

$$f_\eta(x,t;x_0)=(g*_t\eta)(x,t;x_0) \tag{4.17}$$

when $t>\varepsilon$. In the time domain, f_η has the convolution representation

$$f_\eta(x,\cdot;x_0)=\int_{S^2}\eta*m(\omega,\cdot;x_0)*v(x,\cdot;\omega)\,dS(\omega).$$

If $x_0\notin\operatorname{supp} q$, the scattering amplitude determines m.

[10] Because of our definition of the Fourier transform, the sign in front of the argument of the support function must be negative.

Proof. The first statement is clear from Theorem 4.2: for $\varphi \in C_0^\infty(\mathbb{R}^3)$ and $\psi \in C_0^\infty(\mathbb{R})$,

$$\begin{aligned}\langle \mathcal{F}^{-1}(\hat{f}_\eta), \varphi \otimes \psi \rangle &= \langle \hat{f}\hat{\eta}, \varphi \otimes \mathcal{F}^{-1}\psi \rangle \\ &= \langle \hat{f}, \varphi \otimes \left(2\pi(\mathcal{F}^{-1}\mathcal{R}\eta)\mathcal{F}^{-1}\psi\right) \rangle \\ &= \langle \hat{f}, \varphi \otimes \mathcal{F}^{-1}(\mathcal{R}\eta * \psi) \rangle \\ &= \langle f, \varphi \otimes (\mathcal{R}\eta * \psi) \rangle \\ &= \langle f *_t \eta, \varphi \otimes \psi \rangle\end{aligned}$$

by the convolution theorem for test functions. That f_η satisfies (4.17) is seen as follows: If $\operatorname{supp}\psi \subset [\varepsilon, \infty)$,

$$\langle f_\eta - g * \eta, \varphi \otimes \psi \rangle = \langle f - g, \varphi \otimes (\mathcal{R}\eta * \psi) \rangle = 0$$

because $\operatorname{supp}(f-g) \subset \mathbb{R}^3 \times (-\infty, 0]$ and $\operatorname{supp}(\mathcal{R}\eta * \psi) \subset \operatorname{supp}\mathcal{R}\eta + \operatorname{supp}\psi \subset (-\varepsilon, \varepsilon) + [\varepsilon, \infty) = (0, \infty)$.

For the convolution representation, write $\hat{v} = \hat{v}_\mathrm{i} + \hat{v}_\mathrm{s}$, where $\hat{v}_\mathrm{i}(x,k;\omega) = e^{ikx\cdot\omega} = \mathcal{F}(\delta(t - x\cdot\omega))$, and consider the terms separately. For the incident part,

$$\begin{aligned}&\left\langle \mathcal{F}^{-1}\left(\hat{\eta}(k)\int_{S^2} \hat{m}(\omega,k;x_0)\, e^{ikx\cdot\omega}\, dS(\omega)\right)(t), \varphi(x,t)\right\rangle_{x,t} \\ &\quad= \left\langle \hat{\eta}(k)\hat{m}(\omega,k;x_0)\, e^{ikx\cdot\omega}, \check{\varphi}(x,k)\right\rangle_{x,k,\omega} \\ &\quad= \int_{S^2}\int_{\mathbb{R}^3}\left\langle \mathcal{F}(\eta * m)(\omega,k;x_0)e^{ikx\cdot\omega}, \check{\varphi}(x,k)\right\rangle_k dx\,dS(\omega) \\ &\quad= \int_{S^2}\int_{\mathbb{R}^3}\langle(\eta * m)(\omega,t;x_0) *_t \delta(t - x\cdot\omega), \varphi(x,t)\rangle_t\, dx\,dS(\omega) \\ &\quad= \left\langle \int_{S^2}(\eta * m)(\omega,t;x_0) *_t v_\mathrm{i}(x,t;\omega)\, dS(\omega), \varphi(x,t)\right\rangle_{x,t}\end{aligned}$$

since $\delta(\cdot - x\cdot\omega) \in \mathcal{E}'(\mathbb{R})$. For the scattered part, interpret the convolution as $L^1(\mathbb{R},\mathbb{C}) \times L^2(\mathbb{R}, L^2_{-\delta}(\mathbb{R}^3)) \to L^2(\mathbb{R}, L^2_{-\delta}(\mathbb{R}^3))$, since

$$\sup_{\omega\in S^2, x_0\in K}\|\eta * m(\omega,\cdot;x_0)\|_{L^1(\mathbb{R})} \le \sup_{\omega\in S^2, x_0\in K}\|\eta * m(\omega,\cdot;x_0)\|_{L^\infty_2(\mathbb{R})}\|w^{-2}\|_{L^1(\mathbb{R})} < \infty$$

by Lemma 4.5, and

$$\begin{aligned}\|\hat{v}_\mathrm{s}(\cdot,k;\omega)\|_{L^2_{-\delta}} &= \left\|-(I + \mathcal{G}_k Q)^{-1}\mathcal{G}_k q\hat{v}_\mathrm{i}(\cdot,k;\omega)\right\|_{L^2_{-\delta}} \\ &\le \left\|(I + \mathcal{G}_k Q)^{-1}\right\|_{\mathcal{L}(L^2_{-\delta})}\|\mathcal{G}_k\|_{\mathcal{L}(L^2_\delta, L^2_{-\delta})}\|q\hat{v}_\mathrm{i}(\cdot,k;\omega)\|_{L^2_\delta} \le \frac{C}{w(k)}\end{aligned}$$

by Lemma 2.4, and thus by Plancerel's theorem,

$$\|v_\mathrm{s}(\cdot,\cdot;\omega)\|^2_{L^2(\mathbb{R},L^2_{-\delta})} = \|\hat{v}_\mathrm{s}(\cdot,\cdot;\omega)\|^2_{L^2(\mathbb{R},L^2_{-\delta})} \le \int_{-\infty}^{\infty}\|\hat{v}_\mathrm{s}(\cdot,k;\omega)\|^2_{L^2_{-\delta}}\,dk \qquad (4.18)$$

is bounded uniformly for all $\omega \in S^2$. The vector valued convolution theorem [ABHN01, Section 1.8] can be applied, approximating with $L^1(\mathbb{R}, L^2_{-\delta}(\mathbb{R}^3)) \cap L^2(\mathbb{R}, L^2_{-\delta}(\mathbb{R}^3))$ functions, which are dense.

The final statement follows as in Theorem 4.2. □

4.2 Scattering control

Consider waves excited by time-dependent sources in a bounded set $A \subset \mathbb{R}^3$. Since waves travel at unit speed, they will stay supported in the union of light cones with vertices in A. This means that in a finite length of time T, the waves will have travelled at most the distance T away from A. If the sources are "turned on" at time zero, the waves excited by them will be supported at time $T > 0$ in the set of points that are no further than distance T from A. This set is called the domain of influence of A in time T and denoted by A_T.

We shall now show that the time derivatives of waves excited by smooth sources in a bounded domain A actually form a dense subset of all possible waves supported in the domain of influence. In other words, the domain of influence can be approximately controlled from A. [KKL01, Section 3.3]

The results in this section are presented for the three-dimensional case, but the proofs apply without change to any number of dimensions in which the necessary ingredients are available.

Theorem 4.9 (Interior control). *Assume that $q \in C_0^1(\mathbb{R}^3, \mathbb{R})$ is such that the operator $I + \mathcal{G}_0 Q$ is injective on $L^2_{-\delta}$ for some $\delta > 5/2$. Let $A \subset \mathbb{R}^3$ be a bounded set with a C^2 boundary, and let $T > 0$. Write $A_T := \{x \in \mathbb{R}^3 \,|\, d(x,A) \leq T\}$. Let g be the causal Green's function defined by (4.1)–(4.2), and $g_{\eta_\varepsilon}(x,\cdot;x_0) = g(x,\cdot;x_0) * \eta_\varepsilon$ with $\eta \in C_0^\infty((-1,1),[0,\infty))$ and $\eta_\varepsilon(t) = \eta(t/\varepsilon)/\varepsilon$. Then*

$$\Big\{ \partial_t u^{H,\varepsilon}(\cdot,T) \,\Big|\, u^{H,\varepsilon}(x,\cdot) = \int_A H(y,\cdot) * g_{\eta_\varepsilon}(x,\cdot;y)\,dy,$$

$$H \in C_0^\infty(A \times (0,T)),\, 0 < \varepsilon < \operatorname{dist}(\pi_t \operatorname{supp} H, \{0,T\}) \Big\} \tag{4.19}$$

is a dense subset of $L^2(A_T)$.

Proof. The inclusion is clear from the finite speed of wave propagation, since $u^{H,\varepsilon}$ solves

$$\begin{aligned} \left[\partial_t^2 - \Delta + q(x)\right] u^{H,\varepsilon}(x,t) &= (H * \eta_\varepsilon)(x,t) =: H_\varepsilon(x,t) \\ u^{H,\varepsilon}(x,0) &= 0 \\ \partial_t u^{H,\varepsilon}(x,0) &= 0. \end{aligned} \tag{4.20}$$

To show density, let $\psi \in L^2(A_T)$ be any function such that

$$\int_{A_T} \psi(x) \partial_t u^{H,\varepsilon}(x,T)\,dx = 0$$

for all H and ε as in (4.19). The statement follows if we show that $\psi = 0$.

Consider the weak solution $e \in C([0,2T], L^2(\mathbb{R}^3)) \cap C^1([0,2T], H^{-1}(\mathbb{R}^3))$ [KKL01, Corollary 2.36] of the dual problem

$$\begin{aligned} \left[\partial_t^2 - \triangle + q(x)\right] e(x,t) &= 0 \\ e(x,T) &= \psi(x) \\ \partial_t e(x,T) &= 0, \end{aligned}$$

i.e.,

$$\begin{aligned} \langle e, (\partial_t^2 - \triangle + q(x))\varphi\rangle &= \langle 0,\varphi\rangle - \langle 0, \varphi(\cdot,T)\rangle + \langle \partial_t e(\cdot,0), \varphi(\cdot,0)\rangle \\ &\quad + \langle -\psi, \partial_t\varphi(\cdot,T)\rangle - \langle e(\cdot,0), \partial_t\varphi(\cdot,0)\rangle \\ &= \langle \psi, \partial_t\varphi(\cdot,T)\rangle + \langle \partial_t e(\cdot,0), \varphi(\cdot,0)\rangle - \langle e(\cdot,0), \partial_t\varphi(\cdot,0)\rangle \end{aligned}$$

for all $\varphi \in H^2(\mathbb{R}^3 \times [0,T])$. In particular, the choice $\varphi = u^{H,\varepsilon}$ yields

$$\langle e, H_\varepsilon\rangle = \langle \psi, \partial_t\varphi(\cdot,T)\rangle + \langle \partial_t e(\cdot,0), 0\rangle - \langle e(\cdot,0), 0\rangle = 0.$$

Now since functions of the form H_ε are dense in $L^2(A\times(0,T))$, we conclude that $e = 0$ in $A\times[0,T]$, and in particular, its Cauchy data vanish on the boundary of the cylinder with base A,

$$e = \partial_n e = 0 \tag{4.21}$$

on $\partial A\times[0,T]$. Since $e(\cdot,T) = 0$ and the plasma wave equation is translation invariant in time and symmetrical with respect to reversal of time, e is antisymmetrical with respect to reflection about the plane $t = T$, *i.e.*, $e(\cdot, 2T-t) = -e(\cdot,t)$. Equation (4.21) therefore holds in $\partial A\times[0,2T]$. Tataru's theorem of unique continuation [Tat95, Thm. 3] thus shows that e vanishes in the set

$$\left\{(x,t) \in \mathbb{R}^3 \times [0,2T] \,\middle|\, \operatorname{dist}(x,A) \le T - |T-t|\right\}$$

(see Figure 6) and in particular,

$$\psi(x) = e(x,T) = 0 \quad \text{when } x \in A_T.$$

□

In Section 4.1, we noted that a solution

$$u^{H,\varepsilon}(\cdot,t) = \int_A\int_0^T H(y,s)g_{\eta_\varepsilon}(\cdot,t-s;y)\,ds\,dy$$

of (4.20), excited by point sources in the set $A\times[0,T]$, can be simulated using incident plane waves: The values of the function

$$v^{H,\varepsilon}(\cdot,t) = \int_A\int_0^T H(y,s)f_{\eta_\varepsilon}(\cdot,t-s;y)\,ds\,dy$$

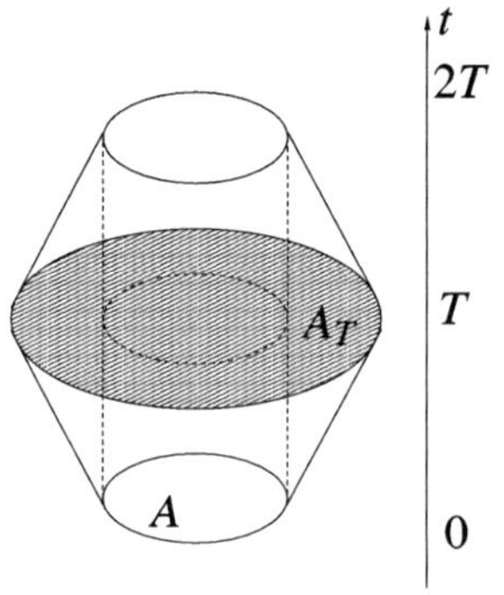

Figure 6: Application of Tataru's uniqueness theorem.

coincide with those of $u^{H,\varepsilon}(\cdot,t)$ for $t \geq \sup(\pi_t \operatorname{supp} H) + \varepsilon$; here π_t denotes the projection onto the time variable. We call the funtion $v^{H,\varepsilon}$ a *wave corresponding to the simulated source* $H_\varepsilon = H * \eta_\varepsilon$.

Theorem 4.9 says that waves corresponding to simulated sources are dense over the domain of influence. In terms of the plane waves used to simulate the sources this yields the following.

Theorem 4.10 (Scattering control). *Assume that $q \in C_0^1(\mathbb{R}^3, \mathbb{R})$ is such that the operator $I + \mathcal{G}_0 Q$ is injective on $L^2_{-\delta}$ for some $\delta > 5/2$. Let $A \subset \mathbb{R}^3$ be a bounded set with C^2 boundary, and let $T > 0$. Write $A_T := \{x \in \mathbb{R}^3 \,|\, d(x,A) \leq T\}$. Let g be the causal Green's function defined by (4.1)–(4.2) and*

$$\hat{m}(\omega,k;x_0) = \frac{k}{2\pi i}\hat{g}^\infty(-\omega,-k;x_0).$$

Let $\eta \in C_0^\infty((-1,1),[0,\infty))$ and $\eta_\varepsilon(t) = \eta(t/\varepsilon)/\varepsilon$. Then

$$\begin{aligned}\Big\{\partial_t v^{H,\varepsilon}|_{A_T}(\cdot,T) \,\Big|\, & v^{H,\varepsilon}(x,\cdot) = \int_{S^2} M(\cdot,\omega) * v(x,\cdot;\omega)\, dS(\omega)\\ & M(\cdot,\omega) := \int_A \eta * m(\cdot,\omega;y) * H(y,\cdot)\, dy \qquad (4.22)\\ & H \in C_0^\infty(A\times(0,T)), 0 < \varepsilon < \operatorname{dist}(\pi_t \operatorname{supp} H, \{0,T\})\Big\}\end{aligned}$$

is a dense subset of $L^2(A_T)$.

Proof. By Theorem 4.8, we see that when $t \geq T$, each function $u^{H,\varepsilon}$ in (4.19) of Theorem 4.9 can be written in the form

$$\begin{aligned}u^{H,\varepsilon}(x,t) &= \int_A \int_0^T H(y,s) f_{\eta_\varepsilon}(x,t-s;y)\, ds\, dy\\ &= \int_{S^2}\int_{\mathbb{R}}\int_A\int_0^T (\eta * m)(t-p-s,\omega;y) H(y,s) ds\, dy\, v(x,p;\omega)\, dp\, dS(\omega).\end{aligned}$$

For v_i, this integral should be interpreted in the sense of distributions. □

5 Solving the inverse problem

The scattering control results of the previous section can now be combined with the Blagoveščenskiĭ identity to solve the inverse problem. For any point $x_0 \in \mathbb{R}^3$ and for any time $t_0 \in \mathbb{R}$, scattering control tells us that a wave can be sent in such that its time derivative does not vanish in a neighbourhood of (x_0,t_0). For such a wave u, the potential can be recovered from

$$q(x_0) = \frac{[\triangle - \partial_t^2]\partial_t u(x_0,t_0)}{\partial_t u(x_0,t_0)}, \tag{5.1}$$

almost as in (1.2). The Blagoveščenskiĭ identity (Theorem 3.12) with $s = t = t_0$ allows us to calculate the L^2 norm of $\partial_t u$. Localizing appropriately, this also yields pointwise information about sufficiently smooth $\partial_t u$, which can be inserted into (5.1).

This will now be done in more detail. In order to be able to use the scattering control results of Section 4, we shall assume that the potential q is continuously differentiable and of bounded support. All the results of this section can be generalized without change to any number of dimensions, assuming that the tools are available.

5.1 Localization

The first step in localizing the solutions of the plasma wave equation is to project them onto balls B. We do this as follows using the Blagoveščenskiĭ identity to calculate inner products with a suitable basis of $L^2(B)$, similarly to [KKL01, Section 4.2.6].

Lemma 5.1. *Assume that $q \in C_0^1(\mathbb{R}^3,\mathbb{R})$ is such that the Schrödinger operator $-\triangle + Q$ has no negative L^2 eigenvalues and no resonance at zero. Then given the scattering amplitude and the incident waves u_{i} and v_{i}, we can calculate the inner products*

$$\big(\partial_t u(\cdot,t_0), \partial_t v(\cdot,s_0)\big)_{L^2(B(x_1,T))}$$

for any $x_1 \in \mathbb{R}^3 \setminus \operatorname{supp} q$, $T > 0$ and any s_0, $t_0 \in \mathbb{R}$.

Proof. Assume for simplicity of notation that $t_0 = 0$. Let $R > 0$ be such that the potential q is supported in B_{R-1}. Consider waves corresponding to simulated sources in the set $A = B(x_1,\rho)$ with $x_1 \in S^2(0,R)$ and $\operatorname{dist}(x_1,\operatorname{supp} q) < \rho < T$ (see Figure 7). By the scattering control Theorem 4.10, we see that at time 0, the time derivatives of waves corresponding to simulated sources in A from time $\rho - T$ to time 0 form a dense subset of all square integrable functions over the domain of influence $A_{T-\rho} = B(x_1,T)$; we can apply the sources on the time interval $[\rho - T, 0]$ instead of $[0, T-\rho]$ because the equation is time-invariant.

Since the Blagoveščenskiĭ identity of Theorem 3.12 allows us to calculate the inner products of the time derivatives of these waves, the Gram-Schmidt process yields an orthonormal basis $\{w_j(\cdot;x_1,T)\}_{j=0}^{\infty}$ of $L^2(A_T)$ such that

$$w_j(x;x_1,T) = \partial_t u_j(x,0)$$

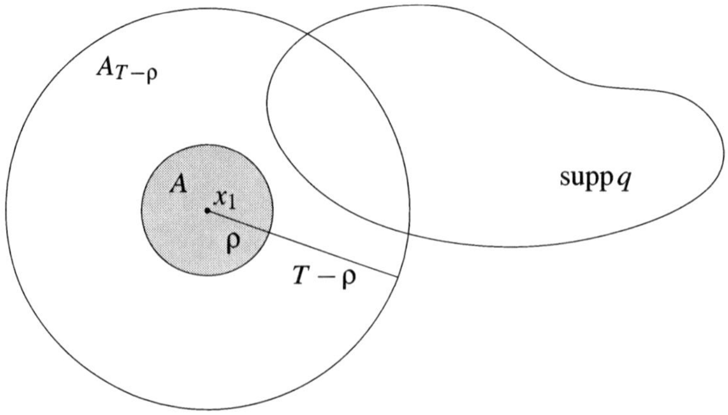

Figure 7: Simulated sources and their domain of influence.

with the functions u_j solving the plasma wave equation. For this reason, this basis is called a *wave basis*. The coordinates of the time derivative of any solution u of (1.1) with respect to this basis can be calculated using the Blagoveščenskiĭ identity:

$$\big(\partial_t u(\cdot,0), w_j(\cdot;x_1,T)\big) = \big(\partial_t u(\cdot,0), \partial_t u_j(\cdot,0)\big).$$

This gives

$$\left[1_{B(x_1,T)}\partial_t u(\cdot,t_0)\right](x) = \sum_{j=0}^{\infty} \big(\partial_t u(\cdot,t_0), w_j(\cdot;x_1,T)\big) w_j(x;x_1,T).$$

The projection of $\partial_t v(\cdot,s_0)$ has an analogous expression. We do not know the values of these functions, since their expressions contain the unknown values of the basis functions $w_j(\cdot;x_1,T)$. However, as the basis is orthonormal, it is possible to calculate the inner products

$$\begin{aligned}\big(\partial_t u(\cdot,t_0), \partial_t v(\cdot,s_0)\big)_{L^2(B(x_1,T))} &= \big(1_{B(x_1,T)}\partial_t u(\cdot,t_0), 1_{B(x_1,T)}\partial_t v(\cdot,s_0)\big)\\ &= \sum_{j=0}^{\infty} \big(\partial_t u(\cdot,t_0), w_j(\cdot;x_1,T)\big)\big(\partial_t v(\cdot,s_0), w_j(\cdot;x_1,T)\big).\end{aligned}$$

Since the potential q is yet to be determined, we do not know *a priori* how large R should have been chosen at the beginning of the present proof. However, we can perform the calculations above for different values of R, and keep increasing its value. When R is so large that the support of the potential is contained in B_{R-1}, the results of our calculations will stop changing. □

Combining projections of this type in an appropriate way, we can take our next step towards finding point values of $\partial_t u$, by projecting onto a small neighbourhood of any point $(x_0,t_0) \in \mathbb{R}^{3+1}$.

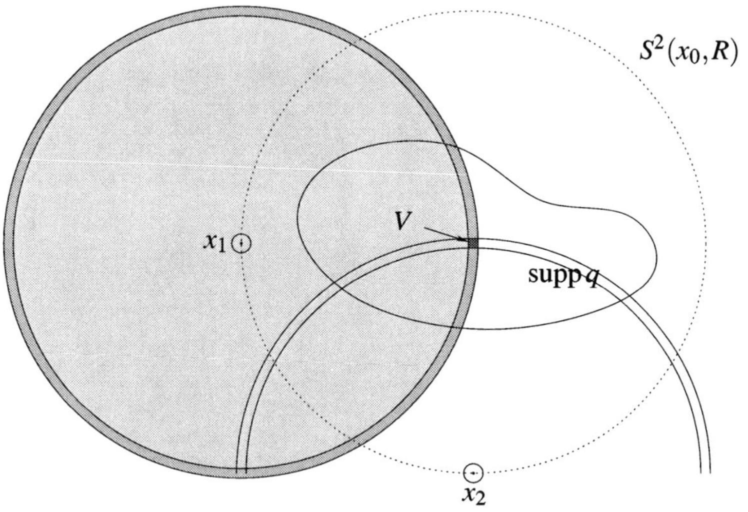

Figure 8: Intersection of the domains of influence of simulated sources around the support of the potential.

Lemma 5.2. *Assume that q is as in Lemma 5.1. Let u and $v \in \tilde{H}$ be solutions of the plasma wave equation (1.1). Then given the scattering amplitude, the incident waves u_{i} and v_{i}, any point $x_0 \in \mathbb{R}^3$, any t_0, $s_0 \in \mathbb{R}$ and any $\delta > 0$ small enough, we can calculate the inner products*

$$\big(\partial_t u(\cdot,t_0), \partial_t v(\cdot,s_0)\big)_{L^2(V(x_0,\delta))} \tag{5.2}$$

where $V(x_0,\delta)$ is a neighbourhood of x_0 such that $B(x_0,c_1\delta) \subset V(x_0,\delta) \subset B(x_0,c_2\delta)$ for some constants c_1, $c_2 \in \mathbb{R}$ independent of δ.

Proof. For $l \in \{1,2,3\}$, set $x_l = x_0 - Re_l$, where R is so large that $\operatorname{supp} q \subset B(x_0,R)$. Let $T_{l\pm} = R \pm \delta/2$ (see Figure 8). Then

$$V := \bigcap_{l=1}^{3} \left[B(x_l,T_{l+}) \setminus B(x_l,T_{l-})\right]$$

is approximately a small cube of side δ with centre at x_0. We can write the projection onto a shell as

$$\begin{aligned}
&\left[1_{B(x_l,T_{l+})\setminus B(x_l,T_{l-})}\partial_t u(\cdot,t_0)\right](x) \\
&\quad = \Big[\left[1 - 1_{B(x_l,T_{l-})}\right] 1_{B(x_l,T_{l+})}\partial_t u(\cdot,t_0)\Big](x) \\
&\quad = \sum_{j=1}^{\infty} \big(\partial_t u(\cdot,t), w_j(\cdot;x_l,T_{l+})\big) \times \\
&\qquad\quad \times \left[w_j(x;x_l,T_{l+}) - \sum_{m=1}^{\infty} \big(w_j(\cdot;x_l,T_{l+}), w_m(\cdot;x_l,T_{l-})\big) w_m(x;x_l,T_{l-})\right],
\end{aligned}$$

where the inner products can be calculated using Lemma 5.1. Repeating this procedure, we find an expression for

$$1_V \partial_t u(\cdot, t_0) = \left[\prod_{l=1}^{3} 1_{B(x_l, T_{l+}) \setminus B(x_l, T_{l-})} \right] \partial_t u(\cdot, t_0)$$

in terms of the wave bases. As in the proof of Lemma 5.1, this yields the inner product (5.2). □

Now in the limit $\delta \to 0$, the set $V(x_0, \delta)$ shrinks nicely to the point x_0, and the projections calculated in the above lemmata give us information about point values:

Lemma 5.3. *Assume that q is as in Lemma 5.1. Let $v = v^{H,\varepsilon}$ be a wave corresponding to a known simulated source as in (4.22). Then given the scattering amplitude, we can calculate the value of $|\partial_t v(x_0, t_0)|$ for all $x_0 \in \mathbb{R}^3$ and $t_0 \in \mathbb{R}$.*

Proof. From Lemma 5.2, we know

$$\frac{\left(\partial_t v(\cdot, t_0), \partial_t v(\cdot, t_0)\right)^{1/2}_{L^2(V(x_0,\delta))}}{m(V(x_0, \delta))} = \frac{\|\partial_t v(\cdot, t_0)\|_{L^2(V(x_0,\delta))}}{m(V(x_0, \delta))} \xrightarrow{\delta \to 0} |\partial_t v(x_0, t_0)| \tag{5.3}$$

for all Lebesgue points of $\partial_t v(\cdot, t_0)$. Now

$$\begin{aligned} \left[\partial_t^2 - \triangle + q(x)\right] \partial_t v^{H,\varepsilon}(x,t) &= \partial_t H * \eta_\varepsilon(x,t) \\ v^{H,\varepsilon}(x,0) &= 0 \\ \partial_t v^{H,\varepsilon}(x,0) &= 0. \end{aligned}$$

Thus as $q \in C_0^1(\mathbb{R}^3)$, $\partial_t v$ is in fact continuous:

$$\partial_t v \in W^{1,\infty}(\mathbb{R}, H^2(\mathbb{R}^3)) \subset C(\mathbb{R}, H^2(\mathbb{R}^3)) \subset C(\mathbb{R}, C(\mathbb{R}^3)) \subset C(\mathbb{R}^{3+1})$$

[Eva98, Thms. 7.2.6 and 5.9.2]. Therefore it makes sense to talk about its point values, and since all points are Lebesgue points for a continuous function, we can find $|\partial_t v(x_0, t_0)|$ for all x_0 and t_0. □

5.2 Recovery of the potential

We are now ready to present the result showing how the Blagoveščenskiĭ identity and scattering control can be used to solve the inverse problem by the localization technique described in Section 5.1.

Theorem 5.4. *Assume that $q \in C_0^1(\mathbb{R}^3, \mathbb{R}^3)$ is such that the Schrödinger operator $-\triangle + Q$ has no negative L^2 eigenvalues and no resonance at zero. Then the scattering amplitude determines the potential q.*

Proof. By scattering control, for any $x_0 \in \mathbb{R}^3$ and $t_0 \in \mathbb{R}$, we can choose a simulated source H and $\varepsilon > 0$ such that the time derivative of the corresponding simulated wave $v = v^{H,\varepsilon}$ does not vanish almost anywhere in a neighbourhood of (x_0,t_0). In this neighbourhood,

$$(\partial_t^2 - \triangle)\partial_t v(x,t) = q(x)\partial_t v(x,t)$$

is continuous as a product of two continuous functions, and we can thus calculate

$$q(x) = \frac{(\partial_t^2 - \triangle)\partial_t v(x,t)}{\partial_t v(x,t)} = \frac{(\partial_t^2 - \triangle)|\partial_t v(x,t)|}{|\partial_t v(x,t)|},$$

provided that u is real valued. In fact, v can be chosen real valued, since for a real valued potential q, the real and imaginary parts of v propagate independently of each other. Thus if $\partial_t v(x_0,t_0) \neq 0$, either $\mathrm{Re}\,\partial_t v$ or $\mathrm{Im}\,\partial_t v$ must be non-zero in a neighbourhood of (x_0,t_0), and we can consider $\mathrm{Re}\,\partial_t v$ or $\mathrm{Im}\,\partial_t v$ instead of v, whichever does not vanish. □

5.3 Prospects

The results in this study lend themselves to generalizations in several different directions: Some of the assumptions made seem artificial and could perhaps be relaxed. On the other hand, our method could be applied in different settings.

One of the main tools that we developed and used, the variant of the Blagoveščenskiĭ identity, was proved under the assumption that the potential and its first derivatives are bounded and decay as $|x|^{-\gamma}$ for $\gamma > 28$. This condition is most probably not sharp. In the proof, a rather long chain of estimates involving many parameters was used, and it would not be surprising if at some point, a looser assumption would suffice. Alternatively, combining the estimates in a different way, or using totally different reasoning, might give the required results under looser assumptions.

The subsequent building blocks that were used for solving the inverse problem were derived under the more restrictive assumption that the potential is of class C_0^1 and that there are no bound states. We therefore concluded only that our method can be used to solve the inverse problem under this assumption. However, the present form of the Blagoveščenskiĭ identity might already prove useful in other applications.

The assumption that $q \in C_0^1$ could perhaps also be relaxed for the scattering control part and when solving the inverse problem. It seems plausible that if a sufficiently quickly decaying potential is smoothly cut off outside a large compact set, the effect will be small. This would enable us to approximate the case of a quickly decaying potential by compactly supported ones.

The assumption about the absence of bound states, *i.e.*, negative eigenvalues of the operator $-\triangle + Q$, might also not be necessary. Under quite general conditions, there are only a finite number of negative eigenvalues, each with a finite dimensional eigenspace. This means that the operator $-\triangle + Q$ is positive definite on a space H' of finite codimension. In this space, scattering can be defined, and the Blagoveščenskiĭ formula derived as in Section 3.3. In Section 5, there is much room for the choice of the functions used in solving the potential q, so it might be possible to choose them from H'.

The principal tools that we used for solving the inverse problem are the Blagoveščenskiĭ identity, developed in Section 3, the scattering control property, developed in Section 4, and the wave bases, developed in Section 5. As these tools are quite geometric in nature, it would be interesting to investigate the possibility of formulating them in a differential geometric setting, and thus use them to solve inverse problems of more general elliptic second order partial differential operators along the lines of [KKL01], and perhaps complementing the active research currently being carried out in the field [AKK$^+$04, KKLM04, JSB99, SB03, SB]. A starting point could be to consider small perturbations to the Euclidean metric, whose effect one would hope to be small enough to permit the present methods to be used. Generalizations to dimensions other than $n = 3$ should also be considered.

An interesting framework for the boundary control method is that of Gaussian beams [BK92a, KK98, KKL01]; this framework could also be fruitful in the study of simulated sources and scattering control. It might also be worthwhile to examine the applicability of the highly localized waves of Section 5.1 to ultrasound surgery [MHK03].

Connections to parabolic and spectral inverse problems [KKLM04, KKL01] might also be worth investigating, as well as the application of ideas presented here to scattering from obstacles and for electromagnetic scattering.

References

[ABHN01] ARENDT, W., C. J. BATTY, M. HIEBER and F. NEUBRANDER: Vector-valued Laplace transforms and Cauchy problems. Monographs in mathematics 96. - Birkhäuser, Basel Boston Berlin, 2001.

[Ada75] ADAMS, R. A.: Sobolev spaces. - Academic Press, New York San Francisco London, 1975.

[Agm75] AGMON, S.: Spectral properties of Schrödinger operators and scattering theory. - Ann. Scuola Norm. Sup. Pisa Cl. Sci. (4) 2(2), 1975, 151–218.

[AKK$^+$04] ANDERSON, M., A. KATSUDA, Y. KURYLEV, M. LASSAS and M. TAYLOR: Boundary regularity for the Ricci equation, geometric convergence, and Gel′fand's inverse boundary problem. - Invent. Math. 158(2), 2004, 261–321.

[Ali84] ALINHAC, S.: Unicité du problème de Cauchy pour des opérateurs du second ordre à symboles réels (Uniqueness of the Cauchy problem for second-order operators with real symbols). - Ann. Inst. Fourier (Grenoble) 34(2), 1984, 89–109. French.

[AP03] ASTALA, K. and L. PÄIVÄRINTA: Calderón's inverse conductivity problem in the plane, 2003. To appear in Annals of Mathematics.

[Bal72] BALANIS, G. N.: The plasma inverse problem. - J. Mathematical Phys. 13, 1972, 1001–1005.

[BB88] BELISHEV, M. and A. BLAGOVEŠČENSKIĬ: Direct method for solving a nonstationary inverse problem for the wave equation. In: Ill-posed problems of mathematical physics and analysis. - Krasnoyarsk, 1988, pp. 43–49. Russian.

[Bel90] BELISHEV, M.: Wave bases in multidimensional inverse problems. - Math. USSR Sbornik 67(1), 1990, 23–42.

[Bel97] BELISHEV, M.: Boundary control in reconstruction of manifolds and metrics (the BC method). - Inverse Problems 13(5), 1997, R1–R45.

[BK92a] BELISHEV, M. I. and A. P. KACHALOV: Boundary control and quasiphotons in a problem of the reconstruction of a Riemannian manifold from dynamic data. - Zap. Nauchn. Sem. S.-Peterburg. Otdel. Mat. Inst. Steklov. (POMI) 203, Mat. Voprosy Teor. Rasprostr. Voln. 22, 1992, 21–50, 174. Russian, English translation in J. Math. Sci. 79(4), 1996, 1172–1190.

[BK92b] BELISHEV, M. I. and Y. V. KURYLEV: To the reconstruction of a Riemannian manifold via its spectral data (BC-method). - Comm. Partial Differential Equations 17(5–6), 1992, 767–804.

[BKS04] BINGHAM, K., Y. V. KURYLEV and E. SOMERSALO (eds.): New analytic and geometric methods in inverse problems. Lectures given at the EMS summer school and conference held in Edinburgh, Scotland 2000. - Springer-Verlag, Berlin Heidelberg New York, 2004.

[Bla71] BLAGOVEŠČENSKIĬ, A. S.: The nonselfadjoint inverse matrix boundary problem for a hyperbolic differential equation. In: Problems of mathematical physics, No. 5: Spectral theory. - Izdat. Leningrad. Univ., Leningrad, 1971, pp. 38–62. Russian.

[Bud61] BUDDEN, K. G.: Radio waves in the ionosphere. - Cambridge University Press, London, 1961.

[Cal80] CALDERÓN, A.-P.: On an inverse boundary value problem. In: Seminar on Numerical Analysis and its Applications to Continuum Physics (Rio de Janeiro, 1980). - Soc. Brasil. Mat., Rio de Janeiro, 1980, pp. 65–73.

[CK98] COLTON, D. and R. KRESS: Inverse acoustic and electromagnetic scattering theory. - Springer-Verlag, Berlin Heidelberg New York, second edition, 1998.

[DR85] DEFACIO, B. and J. H. ROSE: Inverse-scattering theory for the non-spherically-symmetric three-dimensional plasma wave equation. - Physical Review A 31(2), 1985, 897–902.

[ER95] ESKIN, G. and J. RALSTON: Inverse scattering problem for the Schrödinger equation with magnetic potential at a fixed energy. - Comm. Math. Phys. 173(1), 1995, 199–224.

[Eva98] EVANS, L. C.: Partial differential equations. Graduate studies in mathematics 19. - American Mathematical Society, Providence, Rhode Island, 1998.

[Fad56] FADDEEV, L. D.: Uniqueness of solution of the inverse scattering problem. - Vestnik Leningrad. Univ. 11(7), 1956, 126–130.

[Fri80] FRIEDLANDER, F. G.: Radiation fields and hyperbolic scattering theory. - Math. Proc. Cambridge Philos. Soc. 88(3), 1980, 483–515.

[Fri01] FRIEDLANDER, F. G.: Notes on the wave equation on asymptotically Euclidean manifolds. - J. Funct. Anal. 184(1), 2001, 1–18.

[GY99] GÂTEL, Y. and D. YAFAEV: On solutions of the Schrödinger equation with radiation conditions at infinity: the long-range case. - Ann. Inst. Fourier (Grenoble) 49(5), 1999, 1581–1602.

[Hel99] HELGASON, S.: The Radon transform. Progress in mathematics 5. - Birkhäuser, Boston, second edition, 1999.

[Hör90] HÖRMANDER, L.: The analysis of linear partial differential operators I. Distribution theory and Fourier analysis. Springer study edition. - Springer-Verlag, Berlin, second edition, 1990.

[JA79] JORDAN, A. K. and S. AHN: Inverse scattering theory and profile reconstruction. - Proceedings of the Institution of Electrical Engineers 126(10), 1979, 945–950.

[JSB99] JOSHI, M. S. and A. SÁ BARRETO: Recovering asymptotics of metrics from fixed energy scattering data. - Invent. Math. 137(1), 1999, 127–143.

[Kat59] KATO, T.: Growth properties of solutions of the reduced wave equation with a variable coefficient. - Comm. Pure Appl Math. 12, 1959, 403–425.

[Kat95] KATO, T.: Perturbation theory for linear operators. Classics in mathematics. - Springer-Verlag, Berlin, 1995. Reprint of the 1980 edition.

[KK98] KATCHALOV, A. and Y. KURYLEV: Multidimensional inverse problem with incomplete boundary spectral data. - Comm. Partial Differential Equations 23(1–2), 1998, 55–95.

[KKL01] KATCHALOV, A., Y. KURYLEV and M. LASSAS: Inverse boundary spectral problems. Monographs and surveys in pure and applied mathematics 123. - Chapman & Hall/CRC, Boca Raton London New York Washington, D.C., 2001.

[KKLM04] KATCHALOV, A., Y. KURYLEV, M. LASSAS and N. MANDACHE: Equivalence of time-domain inverse problems and boundary spectral problems. - Inverse Problems 20(2), 2004, 419–436.

[LP67] LAX, P. D. and R. S. PHILLIPS: Scattering theory. - Academic Press, 1967.

[Mel94] MELROSE, R. B.: Spectral and scattering theory for the Laplacian on asymptotically Euclidian spaces. In: Spectral and scattering theory (Sanda, 1992), Lecture notes in pure and appl. math. 161. - Dekker, New York, 1994, pp. 85–130.

[Mel95] MELROSE, R. B.: Geometric scattering theory. Stanford lectures. - Cambridge University Press, Cambridge, 1995.

[MF53] MORSE, P. M. and H. FESHBACH: Methods of theoretical physics, part I. - McGraw-Hill, New York, 1953, p. 139.

[MHK03] MALINEN, M., T. HUTTUNEN and J. P. KAIPIO: An optimal control approach for ultrasound induced heating. - Internat. J. Control 76(13), 2003, 1323–1336.

[Nac88] NACHMAN, A. I.: Reconstructions from boundary measurements. - Ann. of Math. (2) 128(3), 1988, 531–576.

[New85] NEWTON, R. G.: Relation between the Schrödinger equation and the plasma wave equation. - Phys. Rev. A (3) 31(5), 1985, 3305–3308.

[NK87] NOVIKOV, R. G. and G. M. KHENKIN: The $\bar{\partial}$-equation in the multidimensional inverse scattering problem. - Uspekhi Mat. Nauk 42(3(255)), 1987, 93–152, 255.

[Nov88] NOVIKOV, R. G.: A multidimensional inverse spectral problem for the equation $-\Delta\psi + (v(x) - Eu(x))\psi = 0$. - Funktsional. Anal. i Prilozhen. 22(4), 1988, 11–22, 96. Russian, English translation in Funct. Anal. Appl. 22(4), 1989, 263–272.

[Nov94] NOVIKOV, R. G.: The inverse scattering problem at fixed energy for the three-dimensional Schrödinger equation with an exponentially decreasing potential. - Comm. Math. Phys. 161(3), 1994, 569–595.

[Päi04] PÄIVÄRINTA, L.: Analytic methods for inverse scattering theory. In: Bingham et al. [BKS04], 2004, pp. 165–185.

[Phi82] PHILLIPS, R. S.: Scattering theory for the wave equation with a short range perturbation. - Indiana University Mathematics Journal 31(5), 1982, 609–639.

[Pot01] POTTHAST, R.: Point sources and multipoles in inverse scattering theory. Research notes in mathematics 427. - Chapman & Hall/CRC, Boca Raton, FL, 2001.

[PS02] PIKE, R. and P. SABATIER (eds.): Scattering. Scattering and inverse scattering in pure and applied science. Vol. 1, 2. - Academic Press Inc., San Diego, CA, 2002.

[Ram87] RAMM, A. G.: Completeness of the products of solutions to PDE and uniqueness theorems in inverse scattering. - Inverse Problems 3(4), 1987, L77–L82.

[Ram88] RAMM, A. G.: Recovery of the potential from fixed-energy scattering data. - Inverse Problems 4(3), 1988, 877–886.

[Ram89] RAMM, A. G.: Corrigendum: "Recovery of the potential from fixed-energy scattering data". - Inverse Problems 5(2), 1989, 255.

[RDC85] ROSE, J. H., B. DEFACIO and M. CHENEY: Physical basis of three-dimensional inverse scattering for the plasma wave equation. - J. Opt. Soc. Amer. A 2(11), 1985, 1954–1957.

[RS79] REED, M. and B. SIMON: Methods of modern mathematical physics III: Scattering theory. - Academic Press, New York, 1979.

[RS00] RAZBOROV, A. G. and V. S. SEROV: On the spectrum of the Schrödinger operator with Kato potential. - Differ. Uravn. 36(5), 2000, 689–693, 720. Russian, English translation in Differential Equations 36(5), 2000, 689–693.

[SB] SÁ BARRETO, A.: Radiation fields, scattering and inverse scattering on asymptotically hyperbolic manifolds. To appear in Duke Math. J.

[SB03] SÁ BARRETO, A.: Radiation fields on asymptotically Euclidean manifolds. - Comm. Partial Differential Equations 28(9–10), 2003, 1661–1673.

[SU87] SYLVESTER, J. and G. UHLMANN: A global uniqueness theorem for an inverse boundary value problem. - Ann. of Math. (2) 125(1), 1987, 153–169.

[Tat95] TATARU, D.: Unique continuation for solutions to PDE's; between Hörmander's theorem and Holmgren's theorem. - Communications in Partial Differential Equations 20(5&6), 1995, 855–884.

[Trè67] TRÈVES, F.: Topological vector spaces, distributions and kernels. Pure and applied mathematics 25. - Academic Press, New York, 1967.

Helsinki University of Technology
Insitute of Mathematics
P.O.Box 1100
FIN-02015 TKK, Finland
Kenrick.Bingham@iki.fi